Florian Ion Petrescu

NEW

AIRCRAFT

II

USA 2012

Scientific reviewer:

Dr. Veturia CHIROIU
Honorific member of
Technical Sciences Academy of Romania (ASTR)
PhD supervisor in Mechanical Engineering

Copyright

Title book: New Aircraft II

Author book: Florian Ion Petrescu

ISBN 978-1-4783-5508-3

WELCOME

The Boeing 787 is the new Boeing aircraft. It is currently in its development phase. Designers of this plane is made lot of research for this aircraft should be particularly fuel-efficient through the use of composite materials in the construction of the device and use of new reactors.

It should enable airlines to reduce by nearly 20% in fuel consumption compared to aircraft of this size.

This aircraft are expected to compete in the world of aircraft types and gain the admiration of the public .

You are welcome to read the full book!
The author.

Content

Boeing 787 Dreamliner

Not a lot is known about this new aircraft, It new concept aircraft "bigger" and "faster", it will carry 225 passengers, top speed between (0.95 and 0.90) Mach, The Sonic Cruiser flies at 45,000 ft"Flying higher offers a smoother ride and gets you above most traffic".

The Sonic Cruiser was proposed shortly after the launch of the Airbus A380 by rival Airbus. the Sonic Cruiser promised 20% faster speed than conventional airplanes,without the noise pollution caused by supersonic .

With this beautifull design the sonic cruiser It very cool airplane in the world of aircraft civil.

Boeing 787 Dreamliner

The Boeing 787 Dreamliner is a long-range, mid-size wide-body, twin-engine jet airliner developed by Boeing Commercial Airplanes. It seats 210 to 290 passengers, depending on the variant. Boeing states that it is the company's most fuel-efficient airliner and the world's first major airliner to use composite materials for most of its construction. According to Boeing, the 787 consumes 20% less fuel than the similarly-sized 767. Its distinguishing features include a four-panel windshield, noise-reducing chevrons on its engine nacelles, and a smoother nose contour. The 787 shares a common type rating with the larger 777 twinjet, allowing qualified pilots to operate both models, due to related design features.

The aircraft's initial designation was 7E7, prior to its renaming in January 2005. The first 787 was unveiled in a roll-out ceremony on July 8, 2007, at Boeing's Everett assembly factory, by which time it had reached 677 orders; this is more orders from launch to roll-out than any previous wide-body airliner. By October 2011, the 787 program had logged 873 orders from 57 customers, with ILFC having the largest number on order.

The 787 development and production has involved a large-scale collaboration with numerous suppliers around the globe. Final assembly is at the

Boeing Everett Factory in Everett, Washington. Assembly is also taking place at a new factory in North Charleston, South Carolina. Both sites will deliver 787s to airline customers. Originally planned to enter service in May 2008, the project has suffered from multiple delays. The airliner's maiden flight took place on December 15, 2009, and completed flight testing in mid-2011. Final Federal Aviation Administration and European Aviation Safety Agency certification was received in late August 2011 and the first model was delivered in late September 2011. It entered commercial service on October 26, 2011.

During the late 1990s, Boeing began considering replacement aircraft programs as sales for the 767 and Boeing 747-400 slowed. The company proposed two new aircraft, the 747X, which would have lengthened the 747-400 and improved efficiency, and the Sonic Cruiser, which would have achieved 15% higher speeds (approximately Mach 0.98) while burning fuel at the same rate as the existing 767. Market interest for the 747X was tepid, but the Sonic Cruiser had brighter prospects. Several major airlines in the United States, including Continental Airlines, initially showed enthusiasm for the Sonic Cruiser concept, although they also expressed concerns about the operating cost.

The global airline market was upended by the September 11, 2001 attacks and increased petroleum prices, making airlines more interested in efficiency than speed. The worst-affected airlines, those in the United States, had been considered the most likely customers of the Sonic Cruiser, and thus Boeing officially cancelled the Sonic Cruiser on December 20, 2002. Switching tracks, the company announced an alternative product using Sonic Cruiser technology in a more conventional configuration, the 7E7, on January 29, 2003. The emphasis on a smaller midsize twinjet rather than a large 747-size aircraft represented a shift from hub-and-spoke theory towards the point-to-point theory, in response to analysis of focus groups.

The replacement for the Sonic Cruiser project was dubbed the "7E7" (with a development code name of "Y2"). Technology from the Sonic Cruiser and 7E7 was to be used as part of Boeing's project to replace its entire airliner product line, an endeavor called the Yellowstone Project (of which the 7E7 became the first stage). Early concept images of the 7E7 included rakish cockpit windows, a dropped nose and a distinctive "shark-fin" tail. The "E" was said to stand for various things, such as "efficiency" or "environmentally friendly"; however, in the end, Boeing claimed that it stood merely for "Eight". In July 2003, a public naming competition was held for the 7E7, for which out of 500,000 votes cast online the winning title was Dreamliner. Other names in the pool included eLiner, Global Cruiser and Stratoclimber.

On April 26, 2004, Japanese airline All Nippon Airways became the launch customer for the 7E7 Dreamliner, by announcing a firm order for 50 aircraft with deliveries to begin in late 2008. All Nippon Airways' order was initially specified as 30 787-3, 290–330 seat, one-class domestic aircraft, and 20 787-8, long-haul, 210–250 seat, two-class aircraft for regional international routes such as Tokyo Narita–Beijing. The aircraft would allow All Nippon Airways to open new routes to cities not previously served, such as Denver, Moscow, and New Delhi.

The 787-3 and 787-8 were to be the initial variants, with the 787-9 entering service in 2010.

The 787 was designed to become the first production composite airliner, with the fuselage assembled in one-piece composite barrel sections instead of the multiple aluminum sheets and some 50,000 fasteners used on existing aircraft. Boeing selected two new engine types to power the 787, the General Electric GEnx and Rolls-Royce Trent 1000. Boeing claimed the 787 would be near to 20 percent more fuel-efficient than the 767, with approximately 40 percent of the efficiency gain from the engines, plus gains from aerodynamic improvements, the increased use of lighter-weight composite materials, and advanced systems. The 787-8 and −9 were intended to be certified to 330 minute ETOPS capability.

During the design phase, the 787 underwent extensive wind tunnel testing at Boeing's Transonic Wind Tunnel, QinetiQ's five-meter wind tunnel at Farnborough, UK, and NASA Ames Research Center's wind tunnel, as well as at the French aerodynamics research agency, ONERA. The final styling of the aircraft was more conservative than earlier proposals, with the fin, nose, and cockpit windows changed to a more conventional form. By the end of 2004, customer-announced orders and commitments for the 787 reached 237 aircraft. Boeing initially priced the 787-8 variant at US$120 million, a low figure that surprised the industry. In 2007, the list price was US$146–151.5 million for the 787-3, US$157–167 million for the 787-8 and US$189–200 million for the 787-9.

After stiff competition, Boeing announced on December 16, 2003, that the 787 would be assembled in its factory in Everett, Washington. Instead of building the complete aircraft from the ground up in the traditional manner, final assembly would employ just 800 to 1,200 people to join completed subassemblies and to integrate systems. Boeing assigned its global subcontractors to do more assembly themselves and deliver completed subassemblies to Boeing for final assembly. This approach was intended to result in a leaner and simpler assembly line and lower inventory, with pre-installed systems reducing final assembly time by three-quarters to three days.

Subcontracted assemblies included wing manufacture (Mitsubishi Heavy Industries, Japan, central wing box) horizontal stabilizers (Alenia Aeronautica, Italy; Korea Aerospace Industries, South Korea); fuselage sections (Global Aeronautica, Italy; Boeing, North Charleston, USA; Kawasaki Heavy Industries, Japan; Spirit AeroSystems, Wichita, USA; Korean Air, South Korea); passenger doors (Latécoère, France); cargo doors, access doors, and crew escape door (Saab AB, Sweden); software development (HCL Enterprise India); floor beams (TAL Manufacturing Solutions Limited, India); wiring (Labinal, France); wing-tips, flap support fairings, wheel well bulkhead, and longerons (Korean Air, South Korea); landing gear (Messier-Dowty, UK/France); and power distribution and management systems, air conditioning packs (Hamilton Sundstrand, Connecticut, USA). Boeing is considering bringing construction of the 787-9 tail in house; the tail of the 787-8 is currently made by Alenia.

To speed delivery of the 787's major components, Boeing modified four used 747-400s into 747 Dreamlifters to transport 787 wings, fuselage sections, and other smaller parts. Japanese industrial participation was very important to the project, with a 35% work share, the first time Japanese firms had taken a lead role in mass production of Boeing airliner wings, and many of the subcontractors supported and funded by the Japanese government. On April 26, 2006, Japanese manufacturer Toray Industries and Boeing announced a production agreement involving US$6 billion worth of carbon fiber, extending a 2004 contract and aimed at easing production concerns.

While Boeing had been working to trim excess weight since assembly of the first airframe began, common for new aircraft in development, the company stated in December 2006 that the first six 787s were overweight, with the first aircraft expected to be 5,000 lb (2,300 kg) heavier than specified. According to International Lease Finance Corporation's (ILFC) Steven Udvar-Hazy, the 787-9's operating empty weight was around 14,000 lb (6,400 kg) overweight. The seventh and subsequent aircraft would be the first optimized 787-8s and were expected to meet all goals, with Boeing working on weight reductions. As part of this process, Boeing redesigned some parts and made more use of lighter titanium.

Boeing had originally planned for a first flight by the end of August 2007 and premiered the first 787 at a rollout ceremony on July 8, 2007, which matches the aircraft's designation in the US-style month-day-year format (7/8/07). However, the aircraft's major systems had not been installed at that time, and many parts were attached with temporary non-aerospace fasteners requiring their later replacement with flight fasteners. Although intended to shorten the production process, 787 subcontractors initially had difficulty completing the extra work, because they could not procure the needed parts, perform the subassembly on schedule, or both, leaving remaining assembly work for Boeing to complete as "traveled work".

On September 5, Boeing announced a three-month delay, blaming a shortage of fasteners as well as incomplete software. On October 10, 2007, a second three-month delay to the first flight and a six-month delay to first deliveries was announced due to problems with the foreign and domestic supply chain, including an ongoing fastener shortage, the lack of documentation from overseas suppliers, and continuing delays with the flight guidance software. Less than a week later, Mike Bair, the 787 program manager was replaced. On January 16, 2008, Boeing announced a third three-month delay to the first flight of the 787, citing insufficient progress on "traveled work". On March 28, 2008, in an effort to gain more control over the supply chain, Boeing announced that it planned to buy Vought Aircraft Industries' interest in Global Aeronautica; the company later agreed to also buy Vought's North Charleston, S.C. factory.

On April 9, 2008, Boeing officially announced a fourth delay, shifting the maiden flight to the fourth quarter of 2008, and delaying initial deliveries by around 15 months to the third quarter of 2009. The 787-9 variant was postponed to 2012 and the 787-3 variant was to follow with no firm delivery date. On November 4, 2008, the company announced a fifth delay due to incorrect fastener installation and the Boeing machinists strike, stating that the first test flight would not occur in the fourth quarter of 2008. After assessing the 787 program schedule with its suppliers, Boeing confirmed on December 11, 2008 that the first flight would be delayed until the second quarter of 2009.

Airlines such as United Airlines and Air India have stated they want compensation from Boeing for the delays.

As Boeing worked with its suppliers on early 787 production, the aircraft design had proceeded through a series of test goals. On August 7, 2007, on-time certification of the Rolls-Royce Trent 1000 engine by European and US regulators was received. On August 23, 2007, a crash test involving a vertical drop of a partial composite fuselage section from about 15 ft (4.6 m) onto a 1 in (25 mm)-thick steel plate occurred in Mesa, Arizona; the results matched what Boeing's engineers had predicted, allowing modeling of various crash scenarios using computational analysis instead of further physical tests. While critics had expressed concerns that a composite fuselage could shatter and burn with toxic fumes during crash landings, Boeing's test data indicated no greater toxicity versus conventional metal airframes. The crash test was the third in a series of demonstrations conducted to match FAA requirements, which included additional certification criteria owing to the 787's introduction of wide-scale composite materials.

The alternative GE GEnx-1B engine achieved certification on March 31, 2008. On June 20, 2008, the 787 team achieved "Power On" of the first aircraft, powering and testing the aircraft's electrical supply and distribution systems. A non-flight 787 test airframe was built for static testing, and on September 27, 2008, over a period of nearly two hours, the fuselage was successfully tested at 14.9 psi (102.7 kPa) differential, which is 150 percent of the maximum pressure expected in commercial service (i.e., when flying at maximum cruising altitude). In December 2008, the Federal Aviation Administration (FAA) passed the maintenance program for the 787.

On May 3, 2009, the first test 787 was moved to the flight line following extensive factory-testing, including landing gear swings, systems integration verification, and a total run-through of the first flight. On May 4, 2009, a press report indicated a 10–15% range reduction, about 6,900 nmi (12,800 km) instead of the originally promised 7,700 to 8,200 nmi (14,800–15,700 km), for early aircraft that were about 8% overweight. Substantial redesign work was expected to correct this, which would complicate increases in production rates; Boeing stated the early 787-8s would have a range of almost 8,000 nmi (14,800 km). As a result, some airlines reportedly delayed deliveries of 787s in order to take later planes that may be closer to the original estimates. Boeing expected to have the weight issues addressed by the 21st production model.

On June 15, 2009, during the Paris Air Show, Boeing said that the 787 would make its first flight within two weeks. However, on June 23, 2009, Boeing announced that the first flight is postponed "due to a need to reinforce an area within the side-of-body section of the aircraft". Boeing provided an updated 787 schedule on August 27, 2009, with the first flight planned to occur by the end of 2009 and deliveries to begin at the end of 2010. The company expected to write off US$2.5 billion because it considered the first three Dreamliners built unsellable and suitable only for flight tests. On October 28, 2009, Boeing announced the selection of Charleston, SC as the site for a second 787 production line, after soliciting bids from multiple states including Washington. On December 12, 2009, the first 787 completed high speed taxi tests, the last major step before flight.

On December 15, 2009, Boeing conducted the Dreamliner's maiden flight with the first 787-8, originating from Snohomish County Airport in Everett, Washington at 10:27 am PST, and landing at Boeing Field in King County, Washington at 1:35 pm PST.

Originally scheduled for four hours, the test flight was shortened to three hours because of bad weather. Boeing's schedule called for a 9-month flight test campaign (later revised to 8.5 months).

The company's previous major aircraft, the 777, took 11 months with nine aircraft, partly to demonstrate 180-min ETOPS, one of its main features.

The 787 flight test program was composed of 6 aircraft, ZA001 through ZA006, four with Rolls-Royce Trent 1000 engines and two with GE GEnx-1B64 engines. The second 787, ZA002 in All Nippon Airways livery, flew to Boeing Field on December 22, 2009 to join the flight test program; the third 787, ZA004 joined the test fleet with its first flight on February 24, 2010, followed by ZA003 on March 14, 2010. On March 24, 2010, testing for flutter and ground effects was completed, clearing the aircraft to fly its entire flight envelope.

On March 28, 2010, the 787 completed the ultimate wing load test, which requires that the wings of a fully assembled aircraft be loaded to 150% of design limit load and held for 3 seconds.

The wings were flexed approximately 25 ft (7.6 m) upward during the test. Unlike past aircraft however, the wings were not tested to failure. On April 7, Boeing announced that analysis of the data showed the test was a success.

On April 23, 2010, Boeing delivered the newest 787, ZA003, to the McKinley Climatic Laboratory hangar at Eglin Air Force Base, Florida for extreme weather testing in temperatures ranging from 115 to -45 °F (46 to -43 °C), and prepare it for takeoff at both temperature extremes. Dreamliner ZA005, the fifth 787 and the first with General Electric GEnx engines began ground engine tests in May 2010. ZA005 made its first flight on June 16, 2010 and joined the flight test program. In June 2010, gaps were discovered in the horizontal stabilizers of test aircraft, due to improperly installed shims; all aircraft produced then were to be inspected and repaired. That same month, a 787 experienced an in-flight lightning strike, allowing engineers the opportunity to examine the aircraft's design tolerances. Since composites can have as much as 1,000 times less electrical conductivity than aluminum, Boeing engineers had added conductive material to ameliorate potential risks and to meet FAA requirements. FAA management was also planning to adjust requirements to help the 787 show compliance. Inspections following the 787's first recorded lightning strike showed no damage to the aircraft.

The 787 made its first appearance at an international air show at the Farnborough Airshow, UK on July 18, 2010. On August 2, a Trent 1000 engine suffered a blowout at Rolls-Royce's test facility during ground testing. The failure caused Boeing to reevaluate its timeline for installing Trent 1000 engines, and on August 27, 2010 the manufacturer confirmed that the first delivery to launch customer All Nippon Airways would be delayed until early 2011. That same month, Boeing faced compensation claims from airlines owing to ongoing delivery delays. On September 9, 2010, it was reported that a further two 787s might join the test fleet, making a total of eight flight test aircraft. On September 10, 2010, a partial engine surge or runaway occurred in a Trent engine on ZA001 at Roswell. On October 4, 2010, the sixth 787, ZA006 joined the test program with its first flight.

On November 5, 2010, it was reported that some early 787 deliveries may be delayed, in one case some three months, to allow for rework to address issues found during flight testing. On November 9, 2010, Boeing 787, ZA002 made an emergency landing after smoke and flames were detected in the main cabin during a test flight over Texas. A Boeing spokeswoman said the airliner landed safely and the crew was evacuated after landing at the Laredo International Airport, Texas. The electrical fire caused some systems to fail before landing. Following this incident, Boeing suspended flight testing on November 10, 2010. Ground testing was performed instead. On November 22, 2010, Boeing announced that the in-flight fire can be primarily attributed to foreign object debris (FOD) that was present in the electrical bay. After electrical system and software changes, the 787 resumed company flight testing on December 23, 2010.

On November 5, 2010, it was reported that some early 787 deliveries may be delayed, in one case some three months, to allow for rework to address issues found during flight testing. On November 9, 2010, Boeing 787, ZA002 made an emergency landing after smoke and flames were detected in the main cabin during a test flight over Texas. A Boeing spokeswoman said the airliner landed safely and the crew was evacuated after landing at the Laredo International Airport, Texas. The electrical fire caused some systems to fail before landing. Following this incident, Boeing suspended flight testing on November 10, 2010. Ground testing was performed instead. On November 22, 2010, Boeing announced that the in-flight fire can be primarily attributed to foreign object debris (FOD) that was present in the electrical bay. After electrical system and software changes, the 787 resumed company flight testing on December 23, 2010.

Regulatory certification of the 787 cleared the way for deliveries to begin. With the first deliveries at hand, Boeing began preparations to increase 787 production rates from two to ten aircraft per month over the next two years. Production is taking place at assembly lines in Everett and Charleston. The Charleston site's contributions have been clouded by legal difficulties; on April 20, 2011, the National Labor Relations Board alleged that Boeing's second production line in South Carolina violated two sections of the National Labor Relations Act. This labor dispute ended in December 2011 when the National Labor Relations Board dropped its lawsuit after the Machinists union withdrew its complaint as part of a new contract with Boeing. The first 787 assembled at the South Carolina facility was rolled out on April 27, 2012.

The first 787 was officially delivered to All Nippon Airways on September 25, 2011, at Boeing's facilities in Everett, Washington. A ceremony to mark the occasion was also held the next day. On September 27, the Dreamliner flew to Haneda Airport. The airline took delivery of the second 787 on October 13, 2011.

On October 26, 2011, the 787 flew its first commercial flight from Narita to Hong Kong on All Nippon Airways. The airliner was planned to enter service some three years prior. Tickets for the flight were sold in an online auction, with the highest bidder paying $34,000 for a seat. ANA expects to take delivery of 7 Dreamliners by the end of 2011 and 9 by March 2012. The 787 flew its first commercial long-haul flight on January 21, 2012 from Haneda to Frankfurt on All Nippon Airways.

On December 6, 2011, test aircraft ZA006 (sixth 787 built), powered by General Electric GEnx engines, flew 10,710 nautical miles (19,830 km) non-stop from Boeing Field eastward to Shahjalal International Airport in Dhaka, Bangladesh, setting a new world distance record for aircraft in the 787's weight class, which is between 440,000 pounds (200,000 kg) and 550,000 pounds (250,000 kg). This flight surpassed the previous record of 9,127 nautical miles (16,903 km), set in 2002 by an Airbus A330. The Dreamliner then continued eastbound from Dhaka to return to Boeing Field, setting a world-circling speed record of 42 hours, 27 minutes.

In late 2011, Boeing began a 787 world tour to promote the airliner. It has visited cities in China, Africa, the Middle East, Europe, United States, and others.

Data from ANA reported that the 787 surpasses the 20% fuel burn reduction promised by Boeing as compared to the 767. On the Tokyo-Frankfurt route the fuel savings was 21%. As part of this report the passenger experience was also rated. Nine in ten passengers said it surpassed their expectations and a quarter said they would go out of their way to fly the 787 again. Air quality and cabin pressure met or exceeded expectations for nine in ten passengers, and 92% of passengers reported that the cabin ambience was as good or better than they expected. Higher humidity levels in the cabin met or exceeded expectations for four in five passengers. Four in ten said headroom was better than expected. Finally, the large windows met or surpassed expectations of 90% of passengers. ANA surveyed 800 passengers who flew the 787 from Tokyo to Frankfurt.

Qatar Airways has placed its first 787 on display at the Farnborough Airshow in July 2012. The airline is to officially receive the aircraft in August.

The 787's design features lighter-weight construction. The aircraft is 80% composite by volume. Its materials, listed by weight, are 50% composite, 20% aluminum, 15% titanium, 10% steel, and 5% other. Aluminum is used on wing and tail leading edges, titanium used mainly on engines and fasteners, with steel used in various places.

External features include raked wingtips and engine nacelles with noise-reducing serrated edges. The longest-range 787 variant can fly 8,000 to 8,500 nautical miles (15,000 to 15,700 km), enough to cover the Los Angeles to Bangkok or New York City to Hong Kong routes. It has a cruising airspeed of Mach 0.85 (561 mph, 903 km/h at typical cruise altitudes).

Among 787 flight systems, the most notable contribution to efficiency is the new electrical architecture, which replaces bleed air and hydraulic power sources with electrically powered compressors and pumps, as well as completely eliminating pneumatics and hydraulics from some subsystems (e.g., engine starters or brakes). Another new system is a wing ice protection system that uses electro-thermal heater mats on the wing slats instead of hot bleed air that has been traditionally used. An active gust alleviation system, similar to the system used on the B-2 bomber, improves ride quality during turbulence.

The 787 flight deck features LCD multi-function displays, all of which will use an industry standard GUI widget toolkit (Cockpit Display System Interfaces to User Systems / ARINC 661). The 787 flight deck includes two head-up displays (HUDs) as a standard feature. Like other Boeing airliners, the 787 will use a yoke instead of a side-stick. The future integration of forward looking infrared into the HUD system for thermal sensing so the pilots can "see" through the clouds is under consideration. The Lockheed Martin Orion spacecraft will use a glass cockpit derived from Honeywell International's 787 flight deck systems.

On the 787, Honeywell and Rockwell Collins provide flight control, guidance, and other avionics systems, including standard dual head up guidance systems, while Thales supplies the integrated standby flight display and electrical power conversion system. A version of Ethernet (Avionics Full-Duplex Switched Ethernet (AFDX) / ARINC 664) will be used to transmit data between the flight deck and aircraft systems.

The airplane's control, navigation, and communication systems are networked with the passenger cabin's in-flight internet systems. In January 2008,

Boeing responded to reports about FAA concerns regarding the protection of the 787's computer networks from possible intentional or unintentional passenger access by stating that various hardware and software solutions are employed to protect the airplane systems.

These included air gaps for the physical separation of the networks, and firewalls for their software separation. These measures prevent data transfer from the passenger internet system to the maintenance or navigation systems.

Each 787 contains approximately 35 short tons (32,000 kg) of carbon fiber reinforced plastic (CFRP), made with 23 tons of carbon fiber. Carbon fiber composites have a higher strength-to-weight ratio than traditional aircraft materials, and help make the 787 a lighter aircraft. Composites are used on fuselage, wings, tail, doors, and interior. Boeing had built and tested the first commercial aircraft composite section while studying the proposed Sonic Cruiser nearly five years before; the Bell Boeing V-22 Osprey military transport uses 50% composites, and the company's C-17 transport has over 16,000 lb (7,300 kg) of structural composites.

Carbon fiber, unlike metal, does not visibly show cracks and fatigue, prompting concerns about the safety risks of widespread use of the material; the rival Airbus A350 was later announced to be using composite panels on a frame, a more traditional approach, which its contractors regarded as less risky. In addition, the porous properties of composite materials, which may cause delamination as collected moisture expands with altitude, is a potential issue. Boeing has responded by noting that composites have been used on wings and other passenger aircraft parts for many years without incident, and that special defect detection procedures will be instituted for the 787 to detect any potential hidden damage.

In 2006, Boeing launched the 787 GoldCare program. This is an optional, comprehensive life-cycle management service whereby aircraft in the program are routinely monitored and repaired as needed. This is the first program of its kind from Boeing. Post-sale protection programs are not new, but have usually been offered by third party service centers. Boeing is also designing and testing composite hardware so inspections are mainly visual. This will reduce the need for ultrasonic and other non-visual inspection methods, saving time and money.

The 787 features two engines. These engines use all-electrical bleedless systems, eliminating the superheated air conduits normally used for aircraft power, de-icing, and other functions. As part of its "Quiet Technology Demonstrator 2" project, Boeing adopted several engine noise-reducing technologies for the 787. Among these are a redesigned air inlet containing sound-absorbing materials and redesigned exhaust duct covers whose rims are tipped in a toothed or chevron pattern to allow for quieter mixing of exhaust and outside air. Boeing expects these developments to make the 787 significantly quieter both inside and out. The noise-reducing measures ensure that sounds above 85 decibels do not leave airport boundaries.

The two different engine models compatible with the 787 use a standard electrical interface to allow an aircraft to be fitted with either Rolls-Royce Trent

1000 or General Electric GEnx engines. This aims to save time and cost when changing engine types; while previous aircraft can have engines changed to those of a different manufacturer, the high cost and time required makes it rare. In 2006, Boeing addressed reports of an extended change period by stating that the 787 engine swap was intended to take 24 hours; engine interchangeability, it is reported, makes the 787 a more flexible asset to airlines, allowing them to change easily from one manufacturer's engine to the other if required.

The 787-8 is designed to seat 234 passengers in a three-class setup, 240 in two-class domestic configuration, and 296 passengers in a high-density economy arrangement. Seat rows can be arranged in four to six abreast in first or business (e.g., 1–2–1, 2–2–2), with eight or nine abreast in economy (e.g., 3–2–3, 2–4–2, 3–3–3). Typical seat room ranges from 46 to 61 in (120 to 150 cm) pitch in first, 36 to 39 in (91 to 99 cm) in business, and 32 to 34 in (81 to 86 cm) in economy.

Cabin interior width is approximately 18 feet (550 cm) at armrest, 1 inch (2.5 cm) more than originally planned, and 15 inches (38 cm) more than that of the Airbus A330 and A340, while 5 inches (13 cm) less than the A350 and 16 in (41 cm) less than the 777. Airlines use economy seat widths that range from 16.33 in (41.5 cm) to 20.66 in (52.5 cm), although published seat research studies recommend a minimum 18-inch (46 cm) seat width. 787 economy seats are approximately 17.2 inches (44 cm) for nine-abreast seating, and 19 inches (48 cm) wide for eight-abreast seating arrangements.

Most airlines are selecting the nine-abreast (3–3–3) configuration. Boeing engineers designed the 787 interior to better accommodate persons with mobility, sensory, and cognitive disabilities.

For example, a 56-inch (142 cm) by 57-inch (145 cm) convertible lavatory includes a movable center wall that allows two separate lavatories to become one large, wheelchair-accessible facility.

The 787's cabin windows are larger in area than all other civil air transports in-service or in development, with dimensions of 10.7 by 18.4 in (27 by 47 cm), and a higher eye level so passengers can maintain a view of the horizon. The composite fuselage permits larger windows without the need for structural reinforcement.

Electrochromism-based "auto-dimming" (smart glass) instead of window shades, supplied by PPG Industries, reduce cabin glare while maintaining transparency. The 787's cabin features light-emitting diodes (LEDs), previously an option on Airbus aircraft and the 777, as standard equipment. The LED system is based in three colors instead of fluorescent tubes, allowing the aircraft to be entirely 'bulbless' and have 128 color combinations.

The internal cabin pressure of the 787 is increased to the equivalent of 6,000 feet (1,800 m) altitude instead of the 8,000 feet (2,400 m) on older conventional aircraft. According to Boeing, in a joint study with Oklahoma State University, this will significantly improve passenger comfort. Cabin air pressurization is provided by electrically driven compressors, rather than traditional engine-bleed air, thereby eliminating the need to cool heated air before it enters the cabin.

The cabin's humidity is programmable based on the number of passengers carried, and allows 15% humidity settings instead of the 4% found in previous aircraft.

The composite fuselage avoids the metal fatigue associated with higher cabin pressure, and eliminates the risk of corrosion from higher humidity levels.

The cabin air-conditioning system improves air quality by removing ozone from outside air, and besides standard HEPA filters which remove airborne particles, uses a gaseous filtration system to remove odors, irritants, and gaseous contaminants as well as particulates like viruses, bacteria and allergens.

Variants:

787-8

The 787-8 is the base model of the 787 family, with a length of 186 feet (57 m) and a wingspan of 197 feet (60 m) and a range of 7,650 to 8,200 nautical miles (14,200 to 15,200 km), depending on seating configuration. The 787-8 seats 210 passengers in a three-class configuration. The variant was the first of the 787 line to enter service, entering service in 2011. Boeing is targeting the 787-8 to replace the 767-200ER and 767-300ER, as well as expand into new non-stop markets where larger planes would not be economically viable. Two thirds of 787 orders are for the 787-8.

787-9

The 787-9 will be the first variant of the 787 with a "stretched" or lengthened fuselage, seating 250–290 in three classes with a range of 8,000 to 8,500 nautical miles (14,800 to 15,750 km). This variant differs from the 787-8 in several ways, including structural strengthening, a lengthened fuselage, a higher fuel capacity, a higher maximum take-off weight (MTOW), but with the same wingspan as the 787-8. The targeted entry into service (EIS) date, was originally planned for 2010, but by October 2011 deliveries were scheduled to begin in early 2014. Boeing is targeting the 787-9 to compete with both passenger variants of the Airbus A330 and to replace their own 767-400ER. Like the 787-8, it will also open up new non-stop routes, flying more cargo and fewer passengers more efficiently than the 777-200ER or A340-300/500. The firm configuration was finalized on July 1, 2010.

When first launched, the 787-9 had the same fuel capacity as the other two variants. The design differences meant higher weight and resulted in a slightly shorter range than the 787-8. After further consultation with airlines, design changes were incorporated to add a forward tank to increase its fuel capacity, so it has a longer range and a higher MTOW than the other two variants. Air New Zealand is the launch customer for the 787-9.

787-3

This variant was designed to be a 290-seat (two-class) short-range version of the 787 targeted at high-density flights, with a range of 2,500 to 3,050 nautical miles (4,650 to 5,650 km) when fully loaded. Its design used the same basic fuselage as the 787-8. The wing was derived from the 787-8, with blended winglets replacing raked wingtips. The change decreased the wingspan by roughly 25 feet (7.6 m), allowing the 787-3 to fit into more domestic gates, and in particular in Japan. This model would have been limited in its range by a reduced MTOW of 364,000 lb (165,100 kg).

787-10

Boeing has stated that it is likely to develop another version, the longer 787-10, with seating capacity between 290 and 310. This proposed model is intended to compete with the planned Airbus A350-900. The 787-10 would supersede the 777-200ER in Boeing's current catalog and could also compete against the Airbus A330-300 and A340-300/500. Boeing was having discussions with potential customers about the 787-10 in 2006 and 2007. In March 2006 Mike Bair, the head of the 787 program at the time, stated that "It's not a matter of if, but when we are going to do it ... The 787-10 will be a stretched version of the 787-9 and sacrifice some range to add extra seat and cargo capacity." Boeing has not yet officially launched the −10, but it remains under consideration as of 2012.

Although no date has been set, Boeing expects to build a 787 freighter version, possibly in 10 to 15 years. Boeing is reported to be also considering a 787 variant as a candidate to replace the 747-based VC-25 as Air Force One.

Airbus

Wings That Waggle Could Cut Aircraft Emissions By 20%

The new approach, which promises to dramatically reduce mid-flight drag, uses tiny air powered jets which redirect the air, making it flow sideways back and forth over the wing.

The jets work by the Helmholtz resonance principle - when air is forced into a cavity the pressure increases, which forces air out and sucks it back in again, causing an oscillation – the same phenomenon that happen when blowing over a bottle.

Dr Duncan Lockerby, from the University of Warwick, who is leading the project, said: "This has come as a bit of a surprise to all of us in the aerodynamics community. It was discovered, essentially, by waggling a piece of wing from side to side in a wind tunnel."

"The truth is we're not exactly sure why this technology reduces drag but with the pressure of climate change we can't afford to wait around to find out. So we are pushing ahead with prototypes and have a separate three year project to look more carefully at the physics behind it."

Part of these savings will be made from lighter aircraft plus improvements in engines and fuel efficiencies but drag (friction caused by is also a major factor in fuel consumption during flights.

Engineers have known for some time that tiny ridges known as 'riblets' - like those found on sharks bodies - can reduce skin-friction drag, (a major portion of mid-flight drag), by around 5%. But the new micro-jet system being developed by Dr Lockerby and his colleagues could reduce skin friction drag by up to 40%,

The research, being carried out with scientists at Cardiff, Imperial, Sheffield, and Queen's University Belfast, is still at concept stage although it is hoped the new wings could be ready for trials as early as 2012.

If successful this technology could also have a major impact on the aerodynamic design and fuel consumptions of cars, boats and trains.

ScienceDaily (May 22, 2009) — Wings which redirect air to waggle sideways could cut airline fuel bills by 20% according to research funded by the Engineering and Physical Sciences Research Council (EPSRC) and Airbus.

Airbus SAS is an aircraft manufacturing subsidiary of EADS, a European aerospace company. Based in Blagnac, France, a suburb of Toulouse, and with significant activity across Europe, the company produces approximately half of the world's jet airliners.

Airbus began as a consortium of aerospace manufacturers, Airbus Industrie. Consolidation of European defence and aerospace companies in 1999 and 2000 allowed the establishment of a simplified joint-stock company in 2001, owned by EADS (80%) and BAE Systems (20%). After a protracted sales process BAE sold its shareholding to EADS on 13 October 2006.

Airbus employs around 63,000 people at sixteen sites in four European Union countries: France, Germany, the United Kingdom and Spain. Final assembly production is at Toulouse (France), Hamburg (Germany), Seville (Spain) and, since 2009, Tianjin (People's Republic of China). Airbus has subsidiaries in the United States, Japan, China and India.

The company produced and markets the first commercially viable fly-by-wire airliner, the Airbus A320, and the world's largest airliner, the A380.

Airbus Industrie began as a consortium of European aviation firms to compete with American companies such as Boeing, McDonnell Douglas, and Lockheed.

While many European aircraft were innovative, even the most successful had small production runs. In 1991, Jean Pierson, then CEO and Managing Director of Airbus Industrie, described a number of factors which explained the dominant position of American aircraft manufacturers: the land mass of the United States made air transport the favoured mode of travel; a 1942 Anglo-American agreement entrusted transport aircraft production to the US; and World War II had left America with "a profitable, vigorous, powerful and structured aeronautical industry.

In the mid-1960s, tentative negotiations commenced regarding a European collaborative approach. Individual aircraft companies had already envisaged such a requirement; in 1959 Hawker Siddeley had advertised an "Airbus" version of the Armstrong Whitworth AW.660 Argosy, which would "be able to lift as many as 126 passengers on ultra short routes at a direct operating cost of 2d. per seat mile." However, European aircraft manufacturers were aware of the risks of such a development and began to accept, along with their governments, that collaboration was required to develop such an aircraft and to compete with the more powerful US manufacturers. At the 1965 Paris Air Show major European airlines informally discussed their requirements for a new "airbus" capable of transporting 100 or more passengers over short to medium distances at a low cost. The same year Hawker Siddeley (at the urging of the UK government) teamed with Breguet and Nord to study airbus designs. The Hawker Siddeley/Breguet/Nord groups HBN 100 became the basis for the continuation of the project. By 1966 the partners were Sud Aviation, later Aérospatiale (France), Arbeitsgemeinschaft Airbus, later Deutsche Airbus (Germany) and Hawker Siddeley (UK). A request for funding was made to the three governments in October 1966. On 25 July 1967 the three governments agreed to proceed with the proposal.

In the two years following this agreement, both the British and French governments expressed doubts about the project. The MoU had stated that 75 orders must be achieved by 31 July 1968. The French government threatened to withdraw from the project due to the concern over funding development of the Airbus A300, Concorde and the Dassault Mercure concurrently, but was persuaded otherwise. Having announced its concern at the A300B proposal in December 1968, and fearing it would not recoup its investment due to lack of sales, the British government announced its withdrawal on 10 April 1969. Germany took this opportunity to increase its share of the project to 50%. Given the participation by Hawker Siddeley up to that point, France and Germany were reluctant to take over its wing design. Thus the British company was allowed to continue as a privileged subcontractor. Hawker Siddeley invested GB£35 million in tooling and, requiring more capital, received a GB£35 million loan from the German government.

Airbus Industrie was formally established as a Groupement d'Interet Économique (Economic Interest Group or GIE) on 18 December 1970. It had been formed by a government initiative between France, Germany and the UK that originated in 1967. The name "Airbus" was taken from a non-proprietary term used by the airline industry in the 1960s to refer to a commercial aircraft of a certain size and range, for this term was acceptable to the French linguistically. Aérospatiale and Deutsche Airbus each took a 36.5% share of production work, Hawker Siddeley 20% and Fokker-VFW 7%. Each company would deliver its sections as fully equipped, ready-to-fly items. In October 1971 the Spanish company CASA acquired a 4.2% share of Airbus Industrie, with Aérospatiale and Deutsche Airbus reducing their stakes to 47.9%. In January 1979 British

Aerospace, which had absorbed Hawker Siddeley in 1977, acquired a 20% share of Airbus Industrie. The majority shareholders reduced their shares to 37.9%, while CASA retained its 4.2%.

The Airbus A300 was to be the first aircraft to be developed, manufactured and marketed by Airbus. By early 1967 the "A300" label began to be applied to a proposed 320 seat, twin engined airliner. Following the 1967 tri-government agreement, Roger Béteille was appointed technical director of the A300 development project. Béteille developed a division of labour which would be the basis of Airbus' production for years to come: France would manufacture the cockpit, flight control and the lower centre section of the fuselage; Hawker Siddeley, whose Trident technology had impressed him, was to manufacture the wings; Germany should make the forward and rear fuselage sections, as well as the upper centre section; the Dutch would make the flaps and spoilers; finally Spain (yet to become a full partner) would make the horizontal tailplane. On 26 September 1967 the German, French and British governments signed a Memorandum of Understanding in London which allowed continued development studies. This also confirmed Sud Aviation as the "lead company", that France and the UK would each have a 37.5% workshare with Germany taking 25%, and that Rolls-Royce would manufacture the engines.

In the face of lukewarm support from airlines for a 300+ seat Airbus A300, the partners submitted the A250 proposal, later becoming the A300B, a 250 seat airliner powered by pre-existing engines. This dramatically reduced development costs, as the Rolls-Royce RB207 to be used in the A300 represented a large proportion of the costs. The RB207 had also suffered difficulties and delays, since Rolls-Royce was concentrating its efforts on the development of another jet engine, the RB211, for the Lockheed L-1011 and

Rolls-Royce entering into administration due to bankruptcy in 1971. The A300B was smaller but lighter and more economical than its three-engined American rivals.

In 1972, the A300 made its maiden flight and the first production model, the A300B2 entered service in 1974; though the launch of the A300 was overshadowed by the similarly timed supersonic aircraft Concorde. Initially the success of the consortium was poor, but orders for the aircraft picked up, due in part to the marketing skills used by Airbus CEO Bernard Lathière, targeting airlines in America and Asia. By 1979 the consortium had 256 orders for A300, and Airbus had launched a more advanced aircraft, the A310, in the previous year. It was the launch of the A320 in 1981 that guaranteed the status of Airbus as a major player in the aircraft market – the aircraft had over 400 orders before it first flew, compared to 15 for the A300 in 1972.

The retention of production and engineering assets by the partner companies in effect made Airbus Industrie a sales and marketing company. This arrangement led to inefficiencies due to the inherent conflicts of interest that the four partner companies faced; they were both GIE shareholders of, and subcontractors to, the consortium. The companies collaborated on development of the Airbus range, but guarded the financial details of their own production activities and sought to maximise the transfer prices of their sub-assemblies. It was becoming clear that Airbus was no longer a temporary collaboration to produce a single plane as per its original mission statement, it had become a long term brand for the development of further aircraft. By the late 1980s work had begun on a pair of new medium-sized aircraft, the biggest to be produced at this point under the Airbus name, the Airbus A330 and the Airbus A340.

In the early 1990s the then Airbus CEO Jean Pierson argued that the GIE should be abandoned and Airbus established as a conventional company.

However, the difficulties of integrating and valuing the assets of four companies, as well as legal issues, delayed the initiative. In December 1998, when it was reported that British Aerospace and DASA were close to merging, Aérospatiale paralysed negotiations on the Airbus conversion; the French company feared the combined BAe/DASA, which would own 57.9% of Airbus, would dominate the company and it insisted on a 50/50 split. However, the issue was resolved in January 1999 when BAe abandoned talks with DASA in favour of merging with Marconi Electronic Systems to become BAE Systems.

Then in 2000 three of the four partner companies (DaimlerChrysler Aerospace, successor to Deutsche Airbus; Aérospatiale-Matra, successor to Sud-Aviation; and CASA) merged to form EADS, simplifying the process. EADS now owned Airbus France, Airbus Deutschland and Airbus España, and thus 80% of Airbus Industrie. BAE Systems and EADS transferred their production assets to the new company, Airbus SAS, in return for shareholdings in that company.

In mid-1988 a group of Airbus engineers led by Jean Roeder began working in secret on the development of an ultra-high-capacity airliner (UHCA), both to complete its own range of products and to break the dominance that Boeing had enjoyed in this market segment since the early 1970s with its 747. The project was announced at the 1990 Farnborough Air Show, with the stated goal of 15% lower operating costs than the 747-400. Airbus organised four teams of designers, one from each of its partners (Aérospatiale, DaimlerChrysler Aerospace, British Aerospace, CASA) to propose new technologies for its future aircraft designs. In June 1994 Airbus began developing its own very large airliner, then designated as A3XX. Airbus considered several designs, including an odd side-by-side combination of two fuselages from the Airbus A340, which was Airbus's largest jet at the time. Airbus refined its design, targeting a 15 to 20 percent reduction in operating costs over the existing Boeing 747–400. The A3XX design converged on a double-decker layout that provided more passenger volume than a traditional single-deck design.

Five A380s were built for testing and demonstration purposes. The first A380 was unveiled at a ceremony in Toulouse on 18 January 2005, and its maiden flight took place on 27 April 2005. After successfully landing three hours and 54 minutes later, chief test pilot Jacques Rosay said flying the A380 had been "like handling a bicycle". On 1 December 2005, the A380 achieved its maximum design speed of Mach 0.96. On 10 January 2006, the A380 made its first transatlantic flight to Medellín in Colombia.

On 3 October 2006, CEO Christian Streiff announced that the reason for delay of the Airbus A380 was the use of incompatible software used to design the aircraft. Primarily, the Toulouse assembly plant used the latest version 5 of CATIA (made by Dassault), while the design centre at the Hamburg factory were using the older and incompatible version 4. The result was that the 530 km of cables wiring throughout the aircraft had to be completely redesigned. Although no orders had been cancelled, Airbus still had to pay millions in late-delivery penalties.

The first aircraft delivered was to Singapore Airlines on 15 October 2007 and entered service on 25 October 2007 with an inaugural flight between Singapore and Sydney. Two months later Singapore Airlines CEO Chew Choong Seng said that the A380 was performing better than both the airline and Airbus had anticipated, burning 20% less fuel per passenger than the airline's existing 747-400 fleet. Emirates was the second airline to take delivery of the A380 on 28 July 2008 and started flights between Dubai and New York on 1 August 2008. Qantas followed on 19 September 2008, starting flights between Melbourne and Los Angeles on 20 October 2008.

In 2003, Airbus and the Kaskol Group created an Airbus Engineering centre in Russia, which started with 30 engineers and since has emerged as a model of success for Airbus' globalisation strategy. It was the first engineering facility to open in Europe outside of the company's home countries. Equipped with state-of-the-art communications equipment and linked with Airbus engineering sites in France and Germany, the facility performs extensive work in disciplines such as fuselage structure, stress, system installation and design. In 2011, the centre employs some 200 engineers who have completed over 30 large-scale projects for the A320, the A330/A340 and the A380 programs. Russian engineers also performed more than half of all design work on the A330-200F freighter, with its activity related to fuselage structure design, floor grids installation and junctions design. The centre currently is involved in the A320neo Sharklets design development and numerous design works for the A350 XWB programme.

On 6 April 2006 plans were announced that BAE Systems was to sell its 20% share in Airbus, then "conservatively valued" at €3.5 billion (US$4.17 billion). Analysts suggested the move to make partnerships with U.S. firms more feasible, in both financial and political terms. BAE originally sought to agree on a price with EADS through an informal process. Due to lengthy negotiations and disagreements over price, BAE exercised its put option which saw investment bank Rothschild appointed to give an independent valuation.

In June 2006 Airbus was embroiled significant international controversy over its announcement of further delays in the delivery of its A380. Following the announcement the value of associated stock plunged by up to 25% in a matter of days, although it soon recovered afterwards. Allegations of insider trading on the part of Noël Forgeard, CEO of EADS, its majority corporate parent, promptly followed. The loss of associated value was of grave concern to BAE, press described a "furious row" between BAE and EADS, with BAE believing the announcement was designed to depress the value of its share. A French shareholder group filed a class action lawsuit against EADS for failing to inform investors of the financial implications of the A380 delays while airlines awaiting deliveries demanded compensation. As a result EADS chief Noël Forgeard and Airbus CEO Gustav Humbert announced their resignations on 2 July 2006.

On 2 July 2006 Rothschild valued BAE's stake at £1.9 billion (€2.75 billion), well below the expectation of BAE, analysts, and even EADS. On 5 July BAE appointed independent auditors to investigate how the value of its share of Airbus had fallen from the original estimates to the Rothschild valuation; however in September 2006 BAE agreed the sale of its stake in Airbus to EADS for £1.87 billion (€2.75 billion, $3.53 billion), pending BAE shareholder approval. On 4 October shareholders voted in favour of the sale, leaving Airbus entirely owned by EADS.

On 9 October 2006 Christian Streiff, Humbert's successor, resigned due to differences with parent company EADS over the amount of independence he would be granted in implementing his reorganisation plan for Airbus. He was succeeded by EADS co-CEO Louis Gallois, bringing Airbus under more direct control of its parent company.

On 28 February 2007, CEO Louis Gallois announced the company's restructuring plans. Entitled Power8, the plan would see 10,000 jobs cut over four years; 4,300 in France, 3,700 in Germany, 1,600 in the UK and 400 in Spain. 5,000 of the 10,000 would be at sub contractors. Plants at Saint Nazaire, Varel and Laupheim face sell off or closure, while Meaulte, Nordenham and Filton are "open to investors". As of 16 September 2008 the Laupheim plant has been sold to a Thales-Diehl consortium to form Diehl Aerospace and while the design activities at Filton have been retained, the manufacturing operations have been sold to GKN of the United Kingdom. The announcements resulted in Airbus unions in France and Germany threatening strike action.

At the 2011 Paris Air Show, Airbus received total orders valued at about $72.2 billion for 730 aircraft, representing a new record in the civil aviation industry. The A320neo ("new engine option") model, announced in December 2010, received 667 orders, which, together with previous orders, resulted in a total of 1029 orders within six months of launch date, also a new record.

The Airbus product line started with the A300, the world's first twin-aisle, twin-engined aircraft. A shorter, re-winged, re-engined variant of the A300 is known as the A310. Building on its success, Airbus launched the A320, particularly notable for being the first commercial jet to utilize a fly-by-wire control system. The A320 has been, and continues to be, a great commercial success. The A318 and A319 are shorter derivatives with some of the latter under construction for the corporate business jet market as Airbus Corporate Jets. A stretched version is known as the A321. The A320 family's primary competitor is the Boeing 737 family.

The longer-range widebody products, the twin-jet A330 and the four-engine A340, have efficient wings, enhanced by winglets. The Airbus A340-500 has an operating range of 16,700 kilometres (9,000 nmi), the second longest range of any commercial jet after the Boeing 777-200LR (range of 17,446 km or 9,420 nautical miles). All Airbus aircraft developed since then have cockpit systems similar to the A320, making it easier to train crew. Production of the four-engine A340 was ended in 2011 due to lack of sales compared to its twin-engine counterparts, such as the Boeing 777.

Airbus is studying a replacement for the A320 series, tentatively dubbed NSR, for "New Short-Range aircraft". Those studies indicated a maximum fuel efficiency gain of 9–10% for the NSR. Airbus however opted to enhance the existing A320 design using new winglets and working on aerodynamical improvements. This "A320 Enhanced" should have a fuel efficiency improvement of around 4–5%, shifting the launch of a A320 replacement to 2017–2018.

In 24 September 2009 the COO Fabrice Bregier stated to Le Figaro that the company would need from €800 million to €1 billion over six years to develop the new aircraft generation and preserve the company technological lead from new competitors like C919, scheduled to operate by 2015–2020.

In July 2007, Airbus delivered its last A300 to FedEx, marking the end of the A300/A310 production line. Airbus intends to relocate Toulouse A320 final assembly activity to Hamburg, and A350/A380 production in the opposite direction as part of its Power8 organisation plan begun under ex-CEO Christian Streiff. Airbus supplied replacement parts and service for Concorde until its retirement in 2003.

The main Airbus factory in Toulouse is located next to Toulouse-Blagnac Airport.

In the late 1990s Airbus became increasingly interested in developing and selling to the military aviation market. Expansion in the military aircraft market is desirable as it reduces Airbus' exposure to downturns in the civil aviation industry. It embarked on two main fields of development: aerial refuelling with the Airbus A310 MRTT and the Airbus A330 MRTT, and tactical airlift with the A400M. In January 1999 Airbus established a separate company, Airbus Military SAS, to undertake development and production of a turboprop-powered tactical transport aircraft, the Airbus Military A400M.

The A400M is being developed by several NATO members, Belgium, France, Germany, Luxembourg, Spain, Turkey, and the UK, as an alternative to relying on foreign aircraft for tactical airlift capacity, such as the Ukrainian Antonov An-124 and the American C-130 Hercules.

The A400M project has received several delays; Airbus has threatened to cancel the development unless it receives state subsidies.

Pakistan placed an order for the Airbus A310 MRTT in 2008, which will be a conversion of an existing airframe as the base model A310 is no longer in production.

On 25 February 2008 it was announced that Airbus had won an order for three air refuelling Multi-Role Tanker Transport (MRTT) aircraft, adapted from A330 passenger jets, from the United Arab Emirates. On 1 March 2008 it was announced that a consortium of Airbus and Northrop Grumman had won a $35 billion contract to build the new in-flight refuelling aircraft KC-45A, a US built version of the MRTT, for the USAF.

The decision drew a formal complaint from Boeing, and the KC-X contract was cancelled to begin bidding afresh.

Airbus is in tight competition with Boeing every year for aircraft orders. Though both manufacturers have a broad product range in various segments from single-aisle to wide-body, their aircraft do not always compete head-to-head. Instead they respond with models slightly smaller or bigger than the other in order to plug any holes in demand and achieve a better edge. The A380, for example, is designed to be larger than the 747. The A350XWB competes with the high end of the 787 and the low end of the 777. The A320 is bigger than the 737-700 but smaller than the 737–800. The A321 is bigger than the 737–900 but smaller than the previous 757-200. Airlines see this as a benefit since they get a more complete product range from 100 seats to 500 seats than if both companies offered identical aircraft.

In recent years the Boeing 777 has outsold its Airbus counterparts, which include the A340 family as well as the A330-300. The smaller A330-200 competes with the 767, outselling its Boeing counterpart in recent years. The A380 is anticipated to further reduce sales of the Boeing 747, gaining Airbus a share of the market in very large aircraft, though frequent delays in the A380 programme have caused several customers to consider the refreshed 747–8. Airbus has also proposed the A350 XWB to compete with the fast-selling Boeing 787 Dreamliner, after being under great pressure from airlines to produce a competing model.

There are around 5,102 Airbus aircraft in service, with Airbus managing to win over 50 per cent of aircraft orders in recent years. Airbus products are still outnumbered 3 to 1 by in-service Boeings (there are over 4,500 Boeing 737s alone in service). This however is indicative of historical success – Airbus made a late entry into the modern jet airliner market (1972 vs. 1958 for Boeing).

Airbus won a greater share of orders in 2003 and 2004. In 2005, Airbus achieved 1111 (1055 net) orders, compared to 1029 (net of 1002) for the same year at rival Boeing However, Boeing won 55% of 2005 orders proportioned by value; and in the following year Boeing won more orders by both measures. Airbus in 2006 achieved its second best year ever in its entire 35 year history in terms of the number of orders it received, 824, second only to the previous year. In August 2010, Airbus announced that it was increasing production of A320 airliners, to reach 40 per month by 2012, at a time when Boeing is increasing monthly 737 production from 31.5 to 35 per month.

Boeing has continually protested over "launch aid" and other forms of government aid to Airbus, while Airbus has argued that Boeing receives illegal subsidies through military and research contracts and tax breaks.

In July 2004 former Boeing CEO Harry Stonecipher accused Airbus of abusing a 1992 bilateral EU-US agreement providing for disciplines for large civil aircraft support from governments. Airbus is given reimbursable launch investment (RLI), called "launch aid" by the US, from European governments with the money being paid back with interest plus indefinite royalties, but only if the aircraft is a commercial success. Airbus contends that this system is fully compliant with the 1992 agreement and WTO rules. The agreement allows up to 33 per cent of the programme cost to be met through government loans which are to be fully repaid within 17 years with interest and royalties. These loans are held at a minimum interest rate equal to the cost of government borrowing plus 0.25%, which would be below market rates available to Airbus without government support. Airbus claims that since the signature of the EU-US agreement in 1992, it has repaid European governments more than U.S.$6.7 billion and that this is 40% more than it has received.

Airbus argues that the military contracts awarded to Boeing, the second largest U.S. defence contractor, are in effect a form of subsidy, such as the controversy surrounding the Boeing KC-767 military contracting arrangements. The significant U.S. government support of technology development via NASA also provides significant support to Boeing, as do the large tax breaks offered to Boeing, which some people claim are in violation of the 1992 agreement and WTO rules. In its recent products such as the 787, Boeing has also been offered direct financial support from local and state governments.

In January 2005 the European Union and United States trade representatives, Peter Mandelson and Robert Zoellick respectively, agreed to talks aimed at resolving the increasing tensions. These talks were not successful with the dispute becoming more acrimonious rather than approaching a settlement.

WTO ruled in August 2010 and in May 2011 that Airbus had received improper government subsidies through loans with below market rates from several European countries. In a separate ruling in February 2011, WTO found that Boeing had received local and federal aid in violation of WTO rules.

Airbus has several final assembly lines for different models and markets. These are:

Hamburg, Germany (A320 series)

Seville, Spain (A400M)

Tianjin, China (A320 series).

Toulouse, France (A320, A330, A340, A380)

Airbus, however, has a number of other plants in different European locations, reflecting its foundation as a consortium. An original solution to the problem of moving aircraft parts between the different factories and the assembly plants is the use of "Beluga" specially enlarged jets, capable of carrying entire sections of fuselage of Airbus aircraft. This solution has also been investigated by Boeing, who retrofitted 3 of their 747 aircraft to transport the components of the 787. An exception to this scheme is the A380, whose fuselage and wings are too large for sections to be carried by the Beluga. Large A380 parts are brought by ship to Bordeaux, and then transported to the Toulouse assembly plant by the Itinéraire à Grand Gabarit, a specially enlarged waterway and road route.

Airbus opened an assembly plant in Tianjin, People's Republic of China for its A320 series airliners in 2009. Airbus started constructing a $350 million component manufacturing plant in Harbin, China in July 2009, which will employ 1,000 people. Scheduled to be operated by the end of 2010, the 30,000 square meter plant will manufacture composite parts and assemble composite work-packages for the A350 XWB, A320 families and future Airbus programs. Harbin Aircraft Industry Group Corporation, Hafei Aviation Industry Company Ltd, AviChina Industry & Technology Company and other Chinese partners hold the 80 percent stake of the plant while Airbus control the remaining 20 percent.

North America is an important region to Airbus in terms of both aircraft sales and suppliers. 2,000 of the total of approximately 5,300 Airbus jetliners sold by Airbus around the world, representing every aircraft in its product line from the 107-seat A318 to the 565-passenger A380, are ordered by North American customers. According to Airbus, US contractors, supporting an estimated 120,000 jobs, earned an estimated $5.5 billion (2003) worth of business. For example, one version of the A380 has 51% American content in terms of work share value.

Plans for a Mobile, Alabama aircraft assembly plant were formally announced by Airbus CEO Fabrice Brégier from the Mobile Convention Center on 2 July 2012. The plans include a $600 million factory at the Brookley Aeroplex for the assembly of the A319, A320 and A321 aircraft. It could employ up to 1,000 full-time workers when operational. Construction is scheduled to begin in 2013, with it becoming operable by 2015 and producing up to 50 aircraft per year by 2017.

Airbus has joined Honeywell and JetBlue Airways in an effort to reduce pollution and dependence on oil. They are trying to develop a biofuel that could be used by 2030. The companies think they can almost cover one third of the world's airplane fuel need. A plan to create a biofuel that won't affect food resources is the proposal. Algae is a possible alternative because it absorbs carbon dioxide, and it will not affect food production. However, algae and other vegetation are still just experiments, and algae is expensive to develop. Airbus recently had the first alternative fuel flight. It ran on 60 percent kerosene and 40 percent gas to liquids (GTL) fuel in one engine. It did not cut carbon emissions, but it was free of sulphur emissions. Alternative fuel was able to work properly in Airbus' aeroplane engine, so alternative fuels should not require new aeroplane engines. This flight and the company's long term efforts are considered big strides towards environmentally friendly aeroplanes.

According to Patrick Crawford of the UK's Export Credits Guarantee Department (ECGD), "Historically, the three European Export Credit Agencies that support Airbus have covered about 17 per cent of that company's total sales. In 2009–10, reflecting the increased constraints on bank liquidity across the world, that proportion rose to 33 per cent. ECGD guarantees represented by Airbus deliveries grew to 90 per cent of the value of business underwritten and 83 per cent of numbers of facilities. Nearly 50 per cent of these Airbus deliveries were powered by UK aero-engines (supplied by either Rolls-Royce or IAE)."

"Son of Concorde" to fly London-to-Sydney in 4 hours?

Recent days have seen reports emerge of a successor to Concorde capable of speeds of over 2,485 mph (4,000 km/h) that could fly from London to Sydney in a mere four hours.

Though very little is known for certain, a joint announcement from aerospace giants Boeing and Lockheed Martin along with business-jet specialists Gulfstream is expected at the imminent Farnborough Air Show, suggesting a collaborative effort between the three corporations. NASA is also said to be offering its assistance.

If accurate, the reported speed would make the supersonic jet, said to be called X-54, almost twice as fast as Concorde. Concorde hasn't flown since its retirement in 2003.

Reports assert that the three companies claim to be close to cracking the problem of the sonic boom, with an engineer purportedly describing the sound the new jet would make as "closer to a puff or plop." It'll be interesting to see if the X-54's eventual design bears any similarity to previously touted noise reduction measures such as v-tails and biplane wings.

It's far from clear who this information comes from, or when and where it emerged. However, Gizmag will be on the ground at the Farnborough Airshow, so if an announcement on a spiritual successor to Concorde is made, we'll be sure to bring you what we know.

From: *http://www.gizmag.com/son-of-concorde/23118/*

First manned flight of FanWing aircraft planned for next year

With a traditional airplane, a propellor or jet engine pulls it forward, and lift is created as air subsequently flows over the wings. FanWing aircraft are a little different.

They have a powered horizontal rotary fan along the leading edge of their single wing, which serves to pull air over it, creating lift without the need for speed. Britain's FanWing company has been developing the technology since 1999, and has already had success with radio-controlled proof-of-concept models. This month, however, the company announced that it plans to debut a two-seater piloted FanWing aircraft at the 2013 EAA AirVenture air show in Oshkosh, Wisconsin.

According to the company, due to the fact that little in the way of speed is required to achieve lift, FanWings can take off and land on very short runways. They are also said to be inexpensive and simple to build, maintain and control; are stable and resistant to turbulence; they won't stall at low speeds; and, they're quiet. Should their engine conk out, their glide ratio is reportedly rather low, although they are still capable of performing an auto-rotational landing.

The piloted aircraft is planned to have a rotor measuring 32 feet (10 meters) long by 30 inches (75 cm) wide, with a total body span (including its twin tails) of 46 feet (14 meters). It will weigh 770 pounds (350 kg) empty, and have a maximum take-off weight of 1,300 pounds (600 kg). Its flight speed will be 20 to 70 knots (23 to 80.5 mph), and it should be able to take off within a distance of 50 feet (15 meters). Power will be supplied by a four-stroke Rotax 912 light aircraft motor.

The FanWing concept was first conceived by American inventor Patrick Peebles, who now heads the company. He is developing the piloted aircraft with a number of collaborators, including former BAE Systems Principal Concepts Engineer George Seyfang.

A small-scale static wind tunnel model has already been built and tested, with construction of the flying prototype scheduled to take place from August to November, followed by the first test flights beginning next February. This is definitely one we'll be watching.

Imperial College reports progress with the new FanWing Ultralight Simulation Project, January-May 31st 2006.

The project will establish major flight performance and pilot-control characteristics for what will eventually act as a fire or other application surveillance manned or unmanned Ultralight FanWing aircraft.

Screen shots shown below are from the developing project of final-year Aeronautics student and pilot Oliver Ahad, whose FanWing Ultralight MSc dissertation is supervised by Professor J M R Graham in consultation with FanWing inventor Patrick Peebles. Images are derived from Peebles's specifications and computer graphics by Jon Linney, Open University KMi.

The Simulator used by Imperial College is a MOTUS Flight Simulation Device manufactured by Fidelity Flight Simulation Inc, Pittsburgh, USA.

The simulator can be configured to simulate a large range of standard civil and military aircraft, and also non-standard designs, using the Laminar Research X-Plane Flight Simulation software. The cockpit seating and general instrumentation layout is configured to represent a typical light twin engine aircraft.

Following continued successful development with increased efficiencies of the UAV airframe prototype (see UAV page), the FanWing Company are now taking steps with UK government backing towards finding major investment and partnership for a first manned aircraft Ultralight prototype.

Preliminary presentation designs of the proposed Ultralight FanWing with 3D images provided by the Open University's Knowledge Media Institute, have been funded by a Jumpstart Connect Award to the Company by the UK Government's London Development Agency (LDA)

The design work will be followed by an Ultralight simulation project at Imperial College, London, to establish performance and control characteristics.

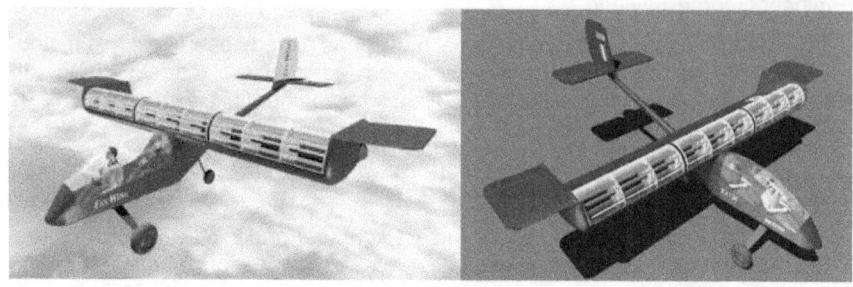

Development of a new manned ultralight FanWing is ongoing and presently planned for a first public flight at Oshkosh 2013.

Recent FanWing development has been based on a collaboration between FanWing inventor Pat Peebles and former BAE Principal Concepts Engineer George Seyfang. Following two series of wind-tunnel tests and analysis from January-June 2011 the newly modified TwinTail prototype was first flight tested June 18th and most recently July 16th 2011.

The twin tails avoid the strong downdraft immediately behind the rotor-wing and simultaneously exploit the updraft from the wingtip vortex. The new development is based on an original suggestion from Seyfang and continues to improve efficiency, speed and stability with the FanWing now capable of reaching speeds of over 70 km/h.

The lower end of the flight envelope (performance at low speeds) was also tested with improved stability at between 20–30 km/h.

Pat Peebles has recently confirmed a new phase of FanWing development with a radically modified design to increase efficiency. The new experiments are based on suggestions by former Principal Future Concept Engineer for BAe Warton, George Seyfang. Seyfang, previously involved in major BAe projects including the Typhoon, the Harrier and the Concorde, has for many years been interested in potential heavy-lift aircraft solutions. Now working independently, he contacted Peebles early this year with suggestions including speed maximization and also increase of efficiency based on the theory of 'Outboard Horizontal Stabilisers'. (The OHS configuration uses the updraft from the wingtip vortex.) Following a series of discussions and a non-disclosure agreement, Peebles and Seyfang have now made development plans for a new private collaborative project.

Preliminary OHS concept feasibility flight tests carried out in July in the UK were already positive and now Peebles reports a successfully sustained test flight on the completed new prototype with swift take-off and improved stability and controllability. Peebles comments: "The onboard data logging will provide details regarding efficiencies, but so far as we can judge from this flight, the new prototype seems to be flight performing beyond predictions."

From: *FANWING,*   *http://www.fanwing.com/news.htm*

SABRE on course to chill SKYLON into orbit
By James Holloway

Reaction Engines has announced that is has successfully tested the key pre-cooler component of its revolutionary SABRE engine crucial to the development of its SKYLON spaceplane. The company claims that craft equipped with SABRE engines will be able to fly to any destination on Earth in under 4 hours, or travel directly into space.

The SABRE engine is capable of operating either as a jet or a rocket, powering aircraft, Reaction Engines claims, at five times the speed of sound within Earth atmosphere, or at 25 times the speed of sound flying directly into orbit. The key to this level of performance is the engine's pre-cooler, which Reaction Engines claims will chill air from over 1000° C (1832° F) to -150° C (-238° F) in under 1/100th of a second.

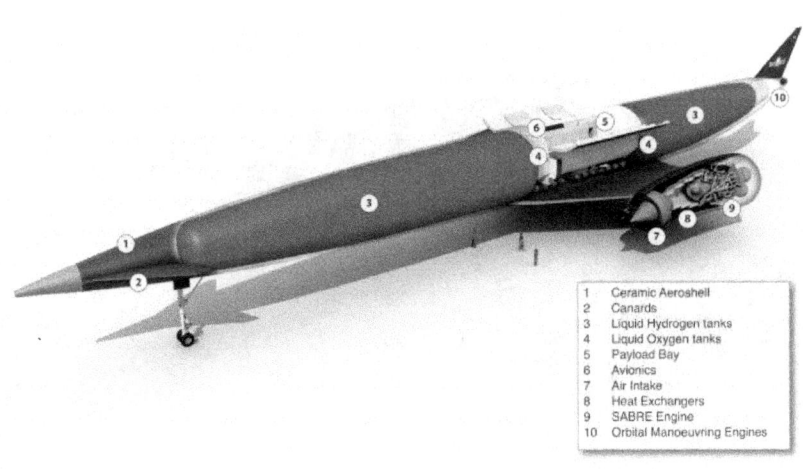

1	Ceramic Aeroshell
2	Canards
3	Liquid Hydrogen tanks
4	Liquid Oxygen tanks
5	Payload Bay
6	Avionics
7	Air Intake
8	Heat Exchangers
9	SABRE Engine
10	Orbital Manoeuvring Engines

The pre-cooler technology was the subject of the latest tests, and Reaction Engines claims its prototype, which is already down to flight-weight, demonstrates the necessary structural integrity, aerodynamic stability and lack of vibration necessary. Though the engine's pre-cooling tests are described as preliminary, further tests commencing in August will push this component to the performance ultimately required.

It has taken a team of more than 30 engineers 22 years to get the SABRE engine where it is today. The cooling assembly used in the engine is on display at the Farnborough Airshow until Sunday July 15.

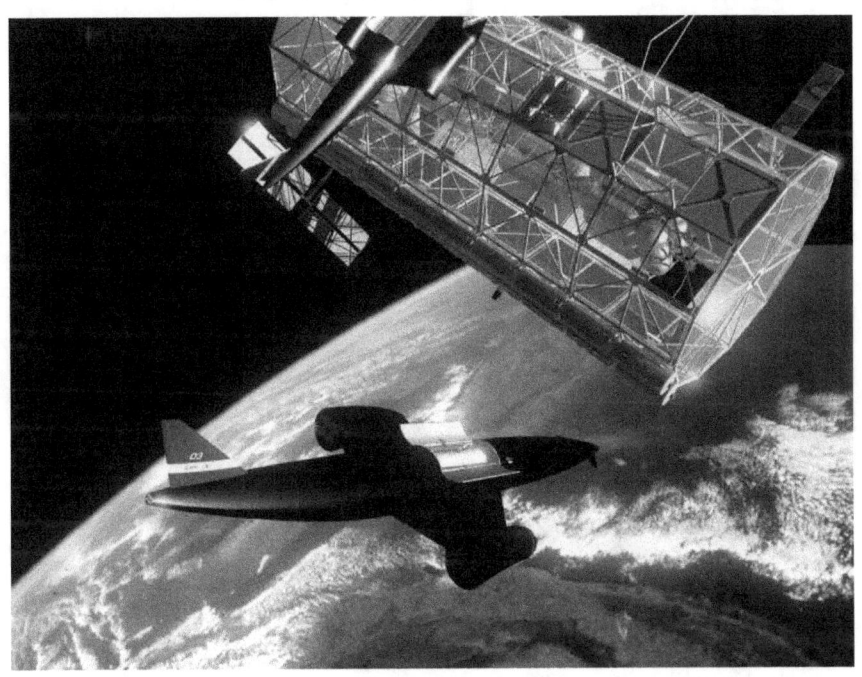

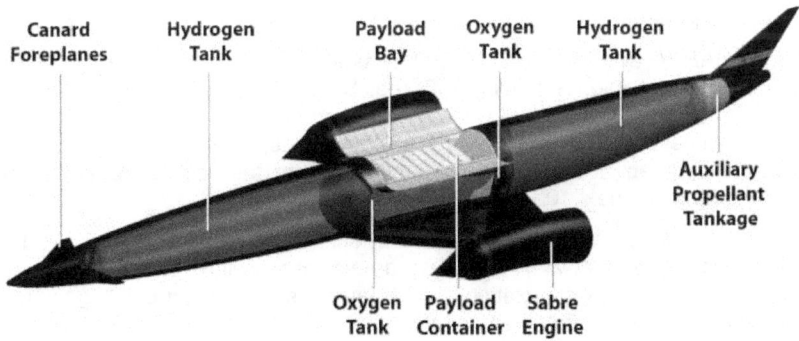

| Canard Foreplanes | Hydrogen Tank | Payload Bay | Oxygen Tank | Hydrogen Tank |
| | | Oxygen Tank | Payload Container | Sabre Engine |

Auxiliary Propellant Tankage

From: *http://www.gizmag.com/sabre-engine/23304*

DARPA funds 100 Year Starship to develop human interstellar flight capabilities

Voyager 1, which is now in the outermost layer of the heliosphere that forms the boundary between the Solar System and interstellar space, is set to be the first man-made object to leave the Solar System. It has taken the car-sized probe over 35 years to reach its current point, but at its current speed of about 3.6 AU (334,640,905 miles) per year it would take over 75,000 years to reach our nearest star, Proxima Centauri. Despite the mind-boggling distances involved, DARPA has just awarded funding to form an organization whose aim is to make human interstellar travel a reality within the next century.

DARPA awarded US$500,000 in seed funding to the Dorothy Jemison Foundation for Excellence to form 100 Year Starship (100YSS), an independent, non-governmental initiative that will call on experts from a variety of fields (artists and entertainers will get a say alongside scientists, engineers and others) to develop the capabilities for human interstellar flight "as soon as possible, and definitely within the next 100 years."

"Yes, it can be done. Our current technology arc is sufficient," said Dr. Mae Jamison, a former NASA astronaut, creator of the winning 100YSS proposal and leader of the new organization. "100 Year Starship is about building the tools we need to travel to another star system in the next 100 years."

The first year of the ambitious project will involve searching for investors, establishing membership opportunities, encouraging public participation in research projects, and developing the visions for interstellar exploration.

A public symposium will also be held in Houston, Texas, from September 13 to 16, 2012, in what will be an annual event "open to scientific papers, engineering challenges, philosophical and socio-cultural considerations, economic incentives, application of space technologies to improve life on Earth, imaginative exploration of the stumbling blocks and opportunities to the stars, and broad public involvement."

The 100YSS initiative will also see the establishment of a scientific research institute called "The Way" that will focus on speculative, long-term science and technology.

"We're embarking on a journey across time and space," says Dr. Jemison. "If my language is dramatic, it is because the project is monumental. This is global aspiration. And each step of the way, its progress will benefit life on Earth. Our team is both invigorated and sobered by the confidence DARPA has in us to start an independent, private initiative to help make interstellar travel a reality."

NASA has launched the Interstellar Boundary Explorer, which will observe the edge of our solar system from a 200,000-mile Earth orbit and determine whether or not we're... err, doomed. Over the next two years, the 23-inch high octagonal craft will study the area of space where solar wind hits the wider galaxy – hopefully it will also find out why the solar wind, which shields us from harmful cosmic rays, has decreased by 25% in the last ten years.

October 21, 2008 NASA has launched the Interstellar Boundary Explorer, which will observe the edge of our solar system from a 200,000-mile Earth orbit and determine whether or not we're, well, doomed. Over the next two years, the 23-inch high octagonal craft will study the area of space where solar wind hits the wider galaxy – hopefully it will also find out why the solar wind, which shields us from harmful cosmic rays, has decreased by 25% in the last ten years.

The solar wind is made up of magnetically charged particles expelled from the sun at one million miles per hour. The particles form a protective shield around our solar system, filtering out 90% of intergalactic radiation. The clashing of the solar wind with external forces was detected by the Voyager crafts, but IBEX was specifically designed to study it.

"No one has seen an image of the interaction at the edge of our solar system where the solar wind collides with interstellar space," said IBEX Principal Investigator David McComas of the Southwest Research Institute in San Antonio. "We know we're going to be surprised. It's a little like getting the first weather satellite images. Prior to that, you had to infer the global weather patterns from a limited number of local weather stations. But with the weather satellite images, you could see the hurricanes forming and the fronts developing and moving across the country."

The Interstellar Boundary Explorer is part of NASA's Small Explorer Program, a series of low-cost data-gathering missions.

Over the course of a year, NASA's Interstellar Boundary Explorer (IBEX) scans the entire sky. During February, its instruments are aligned in the correct direction to intercept atoms that have crossed the boundary from interstellar space into our solar system, become caught by the Sun's gravity and slung around the star. This has now allowed IBEX to capture the most complete glimpse of the material that travels in the galactic wind in the space between star systems. The results indicate this material doesn't look like the same material that makes up our solar system.

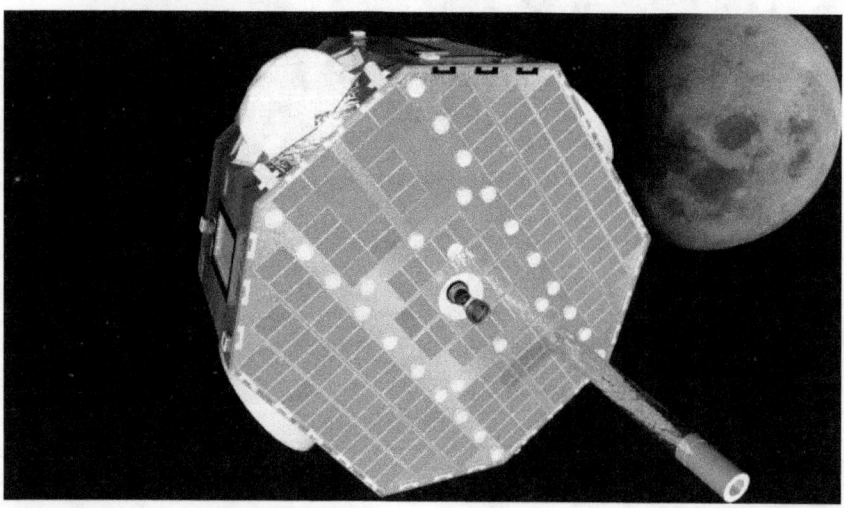

Our solar system is surrounded by a great magnetic bubble known as the heliosphere. Virtually all of the material in the heliosphere emanates from the Sun, which pumps out solar particles that stream to the edge of the solar system and collide with the material in interstellar space at a boundary called the heliosheath. It is this boundary that IBEX, which is currently in a high-altitude sun-oriented elliptical orbit around the Earth, was tasked with mapping when it was launched in October 2008.

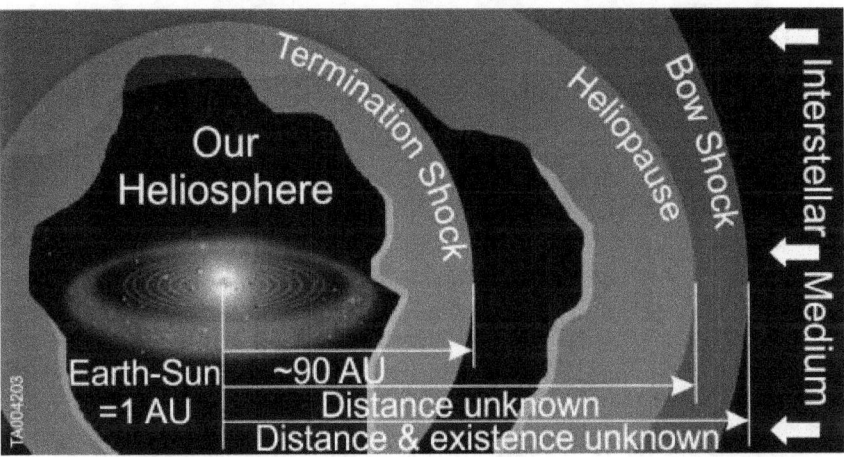

While electrically charged particles traveling through interstellar space on the galactic wind rebound off the heliosheath, neutral particles are able to cross the boundary as if it didn't exist. It is the observation of Energetic neutral atom (ENA) emissions - which are created at the heliosphere by interactions between solar wind particles and interstellar medium particles - by the IBEX and Cassini spacecraft that enabled the first comprehensive sky map of our solar system and its place in the Milky Way galaxy.

Now, NASA says by counting the neutral atoms observed in 2009 and 2010, IBEX has provided clues about how and where our solar system formed, the forces that physically shape our solar system, and even the history of other stars in the Milky Way.

"We've directly measured four separate types of atoms from interstellar space and the composition just doesn't match up with what we see in the solar system," says Eric Christian, mission scientist for IBEX at NASA's Goddard Space Flight Center. "IBEX's observations shed a whole new light on the mysterious zone where the solar system ends and interstellar space begins."

The first direct measurements of hydrogen, helium, oxygen and neon from outside the solar system achieved by IBEX indicate that for every 20 neon atoms in the galactic wind, there are 74 oxygen atoms. However, for every 20 neon atoms in our solar system there are 111 atoms of oxygen. This means that there is more oxygen in any given slice of our solar system than there is in interstellar space.

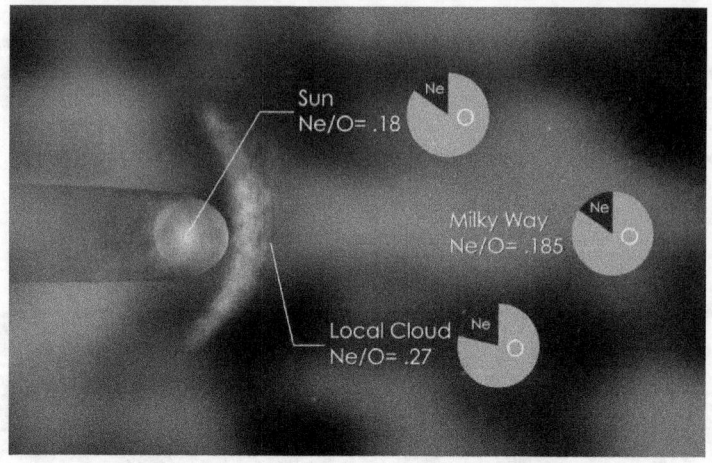

"Our solar system is different than the space right outside it and that suggests two possibilities," says David McComas the principal investigator for IBEX at the Southwest Research Institute in San Antonio, Texas. "Either the solar system evolved in a separate, more oxygen-rich part of the galaxy than where we currently reside or a great deal of critical, life-giving oxygen lies trapped in interstellar dust grains or ices, unable to move freely throughout space." Whatever the answer, NASA says the findings affect scientific models of how our solar system - and life - formed.

Additionally, while hydrogen and helium were initially created in the big bang, the heavier elements of oxygen and neon can only be spread through the galaxy by the supernovae explosions at the end of a giant star's life. NASA says knowing the amounts of such elements in space can help map how the galaxy has evolved and changed over time.

IBEX has also been able to measure the pressure exerted on our heliosphere from interstellar material, which will help scientists determine the size and shape of our solar system as it makes its way through the galaxy. The IBEX measurements also indicate that, although researchers had previously theorized that the solar system may lay at the boundary of the local interstellar cloud in which it resides and may be transitioning into a new region of space, we actually remain fully in the cloud - for the time being anyway.

"Sometime in the next hundred to few thousand years, the blink of an eye on the timescales of the galaxy, our heliosphere should leave the local interstellar cloud and encounter a much different galactic environment," McComas says.

A series of papers examining the IBEX findings appeared in The Astrophysics Journal on January 21, 2012.

"This set of papers provide many of the first direct measurements of the interstellar medium around us," says McComas. "We've been trying to understand our galaxy for a long time, and with all of these observations together, we are taking a major step forward in knowing what the local part of the galaxy is like."

From: *http://www.gizmag.com/darpa-funds-100-year-starship/22662/pictures*

Queen Elizabeth class aircraft carrier

The Queen Elizabeth-class (formerly the CV Future or CVF project) is a class of two aircraft carriers being built for the Royal Navy. HMS Queen Elizabeth is expected to enter service in 2016 and HMS Prince of Wales in 2018.

The contract for the vessels was announced on 25 July 2007 by then Secretary of State for Defence Des Browne, ending several years of delay over cost issues and British naval shipbuilding restructuring; the cost was initially estimated to be £3.9 billion. The contracts were officially signed one year later on 3 July 2008 after the creation of BVT Surface Fleet through the merger of BAE Systems Surface Fleet Solutions and VT Group's VT Shipbuilding which was a requirement of the UK Government.

The vessels will displace about 65,000 metric tons (64,000 long tons), be 280 metres (920 ft) long and have a tailored air group of up to forty aircraft. They will be the largest warships ever to be constructed for the Royal Navy.

The carriers will be completed as originally planned, in a Short Take-Off and Vertical Landing (STOVL) configuration, deploying the Lockheed Martin F-35B. Following the 2010 Strategic Defence and Security Review, the British government had intended to purchase the F-35C carrier version of this aircraft, and adopted plans for Prince of Wales to be built to a Catapult Assisted Take Off Barrier Arrested Recovery (CATOBAR) configuration. After the projected costs of the CATOBAR system rose to around twice the original estimate, the government announced that it would revert to the original design on 10 May 2012.

Under the previous plans, the Royal Navy would operate only one aircraft carrier, routinely equipped with 12 fast jets. However, the Chief of the Defence Staff has subsequently said that the STOVL design, "gives us the ability to operate two carriers if we choose." The final decision will be made at the next major strategic defence review, expected in 2015.

In May 1997, the newly elected Labour government launched the Strategic Defence Review (SDR) which re-evaluated every weapon system (active or in procurement) with the exception of the Eurofighter Typhoon and the Vanguard-class ballistic missile submarines. The report, published in July 1998 identified that aircraft carriers offered the following.

Ability to operate offensive aircraft abroad when foreign basing may be denied.

All required space and infrastructure; where foreign bases are available they are not always available early in a conflict and infrastructure is often lacking.

A coercive and deterrent effect when deployed to a trouble spot.

The report concluded: "the emphasis is now on increased offensive air power, and an ability to operate the largest possible range of aircraft in the widest possible range of roles. When the current carrier force reaches the end of its planned life, we plan to replace it with two larger vessels. Work will now begin to refine our requirements but present thinking suggests that they might be of the order of 30,000–40,000 tonnes and capable of deploying up to fifty aircraft, including helicopters."

The vessels, described as "supercarriers" by the media, legislators and sometimes by the Royal Navy will displace approximately 65,000 t (64,000 long tons) each, over three times the displacement of the current Invincible class. They will be the largest warships ever built in the United Kingdom. The last large carriers proposed for the Royal Navy, the CVA-01 programme, had been cancelled by the Labour government in 1966. In November 2004, giving evidence to the House of Commons Defence Committee, the then First Sea Lord Admiral Sir Alan West explained that the sortie rate and interoperability with the United States Navy were factors in deciding on the size of the carriers and the composition of the carriers' air-wings.

The reason that we have arrived at what we have arrived at is because to do the initial strike package, that deep strike package, we have done really quite detailed calculations and we have come out with the figure of 36 joint strike fighters, and that is what has driven the size of it, and that is to be able to deliver the weight of effort that you need for these operations that we are planning in the future. That is the thing that has made us arrive at that size of deck and that size of ship, to enable that to happen. I think it is something like 75 sorties per day over the five-day period or something like that as well.

I have talked with the CNO (Chief of Naval Operations) in America. He is very keen for us to get these because he sees us slotting in with his carrier groups. For example, in Afghanistan last year they had to call on the French to bail them out with their carrier. He really wants us to have these, but he wants us to have same sort of clout as one of their carriers, which is this figure at 36. He would find that very useful, and really we would mix and match with that.

On 25 January 1999, six companies were invited to tender for the assessment phase of the project – Boeing, British Aerospace, Lockheed Martin, Marconi Electronic Systems, Raytheon and Thomson-CSF. On 23 November 1999, the Ministry of Defence (MoD) awarded detailed assessment studies to two consortia, one led by BAe (renamed BAE Systems on 30 November 1999) and one led by Thomson-CSF (renamed Thales Group in 2000). The brief required up to six designs from each consortium with air-groups of thirty to forty Future Joint Combat Aircraft (FJCA). The contracts were split into phases; the first £5.9 million phase was for design assessment which would form part of the aircraft selection, while the second £23.5 million phase involved "risk reduction on the preferred carrier design option."

On 17 January 2001, the UK signed a Memorandum of Understanding (MoU) with the United States Department of Defense (DoD) for full participation in the Joint Strike Fighter (JSF) programme, confirming the JSF as the FJCA. This gave the UK input into aircraft design and the choice between the Lockheed Martin X-35 and Boeing X-32. On 26 October 2001, the DoD announced that Lockheed Martin had won the JSF contract.

On 30 September 2002, the MoD announced that the Royal Navy and Royal Air Force would operate the STOVL F-35B variant. Also announced was that the carriers would take the form of large, conventional carriers, initially adapted for STOVL operations. The carriers, expected to remain in service for fifty years, were designed for but not with catapults and arrestor wires. The carriers were thus planned to be "future proof", allowing them to operate a generation of CATOBAR aircraft beyond the F-35. Four months later on 30 January 2003, Defence Secretary Geoff Hoon announced that the Thales Group design had won the competition but that BAE Systems would operate as prime contractor.

The contract for the vessels was announced on 25 July 2007 by then Secretary of State for Defence Des Browne, ending several years of delay over cost issues and British naval shipbuilding restructuring; the cost was initially estimated to be £3.9 billion. The contracts were officially signed one year later on 3 July 2008 after the creation of BVT Surface Fleet through the merger of BAE Systems Surface Fleet Solutions and VT Group's VT Shipbuilding which was a requirement of the UK Government.

Then in August 2009, speculation mounted that the UK would drop the F-35B for the F-35C model, which would have meant the carriers being built to operate conventional take off and landing aircraft using the US-designed Electromagnetic Aircraft Launch System (EMALS) catapults.

On 19 October 2010, the government announced the results of its Strategic Defence and Security Review. The review stated that only one carrier was certain to be commissioned; the fate of the other was left undecided. The second ship of the class could be placed in "extended readiness" to provide a continuous single carrier strike capability when the other was in refit, or provide the option to regenerate more quickly to a two carrier strike ability. Alternatively, the second ship could be sold in "cooperation with a close ally to provide continuous carrier-strike capability."

It was also announced that the operational carrier would have catapult and arrestor gear (CATOBAR) installed to accommodate the carrier variant of the Joint Strike Fighter rather than the short-take off and vertical-landing version.

The decision to convert Prince of Wales to CATOBAR was reviewed after the projected costs rose to around double the original estimate. On 10 May 2012 Defence Secretary Philip Hammond announced in Parliament that the government had decided to revert to its predecessor's plans to purchase the F-35B rather than the F-35C, and to complete both aircraft carriers with "ski-jumps" in the STOVL configuration. The total cost of the work that had been done on the conversion to a CATOBAR configuration, and of reverting the design back to the original STOVL configuration, was estimated by Philip Hammond to be "something in the order of £100 million."

The ships' company is 679 rising to 1,600 with air element added. At full displacement they will weigh 65,000 metric tons (64,000 long tons). They have an overall length of 280 metres (920 ft), a width at deck level of 70 metres (230 ft), a height of 56 metres (184 ft), a draught of 11 metres (36 ft) and a range of 10,000 nautical miles (12,000 mi; 19,000 km). Power is supplied by two Rolls-Royce Marine Trent MT30 36 MW (48,000 hp) gas turbine generator units and four Wärtsilä diesel generator sets (two 9 MW (12,000 hp) and two 11 MW (15,000 hp) sets), giving a top speed of 25 knots (46 km/h). The Trents and diesels are the largest ever supplied to the Royal Navy, and together they feed the low-voltage electrical systems as well as the two tandem electric propulsion motors that drive the twin fixed-pitch propellers.

On the flight deck, the equivalent in size of three football pitches, are two small islands instead of a traditional large single island. The forward island is for navigating the ship, while the aft island is for controlling flying operations. Under the flight deck are a further nine decks. The hangar deck measures 155 by 33.5 metres (509 by 109.9 ft) with a height of 6.7 to 10 metres (22 to 33 ft), large enough to accommodate up to twenty fixed and rotary wing aircraft. To transfer aircraft from the hangar to the flight deck, the ships have two large lifts, each of which are capable of lifting two F-35 sized aircraft from the hangar to the flight deck in sixty seconds. The ships' only announced self defence weapons are currently the Phalanx CIWS for airborne threats, with miniguns and 30 mm cannons to counter seaborne threats.

The ship's radars will be the BAE Systems S1850M, the same as fitted to the Type 45 destroyers, for long range wide area search, and the BAE Systems Artisan 3D maritime medium-range radar and a navigation radar. BAE claims the S1850M has a fully automatic detection and track initiation that can track up to 1,000 air targets at a range of around 400 kilometres (250 mi). Artisan can "track a target the size of a snooker ball over 20 kilometres (12 mi) away". (This system will also be fitted to Type 23 frigates, the assault ships HMS Albion, HMS Bulwark and HMS Ocean.) They will also be fitted with the Ultra Electronics Series 2500 Electro Optical System (EOS) and Glide Path Camera (GPC)

Munitions and ammunition handling is accomplished using a highly mechanised weapons handling system (HMWHS). This is a first naval application of a common land based warehouse system. The HMWHS moves palletised munitions from the magazines and weapon preparation areas, along track ways and via several lifts, forward and aft or port and starboard. The tracks can carry a pallet to magazines, the hangar, weapons preparation areas, and the flight deck. In a change from normal procedures the magazines are unmanned, the movement of pallets is controlled from a central location, and manpower is only required when munitions are being initially stored or prepared for use. This system speeds up delivery, and reduces the size of the crew by automation.

Crew facilities will include a cinema, physical fitness areas and four galleys manned by sixty-seven catering staff. There are four large dining areas, the largest with the capacity to serve 960 meals in one hour. There are eleven medical staff for the eight bed medical facility, which includes an operating theatre and a dental surgery.

The vessels are expected to be capable of carrying forty aircraft, a maximum of thirty-six F-35s and four helicopters. Instead of all naval aircraft, depending on the mission a mixture of aircraft could be carried, including up to twelve Royal Air Force Chinook and eight British Army Apache helicopters.

With the retirement of the Harrier GR7/9 in 2010, there are no carrier-capable fixed-wing aircraft available in the Royal Navy or Royal Air Force. Their expected replacement is the Lockheed Martin F-35 Lightning II.

As originally intended, the ships will carry the STOVL version, the F-35B. The aircraft will be flown by pilots from the Fleet Air Arm and the Royal Air Force. The aircraft are expected to begin trials flying from the Queen Elizabeth in 2018 with a carrier air wing fully operational by 2020.

For a period following the 2010 Strategic Defence and Security Review, the government had intended to purchase the F-35C carrier variant and modify one carrier to use the CATOBAR system to launch and recover these aircraft.

This was because the cheaper F-35C variant has a greater range and can carry a larger and more diverse payload than the F-35B. However, on 10 May 2012 Defence Secretary Philip Hammond announced in Parliament that the government had decided to revert to its predecessor's plans to purchase the F-35B rather than the F-35C, and to abandon the completion of Prince of Wales to a CATOBAR configuration.

The reason given was that, "conversion to 'cats and traps' will cost about double what was originally estimated – and would not be delivered until 2023 at the earliest." On the 19 July 2012 the Defence Secretary, Philip Hammond indicated in a speech in the USA that the UK would initially take delivery of 48 f35B for the aircraft carriers and announce at a later date what the final numbers will be.

The AgustaWestland AW101, or Merlin, is a medium-sized multi-role helicopter.

Two versions are in service with the UK armed forces: a utility version, that can carry up to thirty-eight troops or sixteen stretcher patients, and a dedicated anti-submarine warfare variant, with a dipping sonar and sonar-buoys, and a complete electronic warfare suite.

Both versions use a common airframe, with three Rolls-Royce Turbomeca RTM322 engines, their range and endurance using only a two engine cruise option, is 750 nautical miles (1,390 km; 860 mi), or six hours.

However, range can be extended further when the five underfloor fuel tanks are supplemented with auxiliary fuel tanks fitted in the cabin.

Armament depends on mission, but includes anti-ship missiles, torpedoes, three door mounted machine guns, multi-purpose rocket, cannon pods, air to air missiles and air to surface missiles.

The AgustaWestland Lynx Wildcat, is scheduled to enter service with the Royal Navy in 2015.

The Wildcat can be equipped with several mission sensors, which can include: radar, active dipping sonar, electro-optical imaging, electronic surveillance measures and an integrated self defence suite.

The maritime version can be armed with air to surface missiles, torpedoes, depth charges, cannons and heavy machine guns.

The aircraft has a maximum range of 520 nautical miles (960 km; 600 mi) and an endurance of four and a half hours.

The need for a new airborne early warning and control (AEW) platform was identified at an early stage and was an integral part of the next-generation aircraft carrier, and the future carrier-borne aircraft plans. The programme became known as the "Future Organic Airborne Early Warning" (FOAEW), and contracts were placed with BAE / Northrop Grumman and Thales in April 2001. In April 2002, BAE and Northrop Grumman received a follow-on study contract for Phase II of the project, by then renamed Maritime Airborne Surveillance and Control (MASC).

In September 2005, the MASC assessment phase for an AEW aircraft to succeed the Sea King ASaC7 helicopter began. By May 2006, three study contracts were awarded for MASC platform and mission systems options: one to Lockheed Martin UK for a Merlin helicopter fitted with AEW mission systems, another to AgustaWestland who plan to maintain the present Sea King ASaC7 to 2017 and finally to Thales UK to upgrade the Sea King's mission systems.

However, it has been suggested that when the Sea Kings are retired, the MASC role could be undertaken by an AEW variant of the V-22 Osprey. If the carriers had been completed in a CATOBAR configuration, the E-2 Hawkeye could have been another option. This would have had greater speed and range, a more powerful radar than a helicopter-based system, and cross-deck interoperability with the United States and France.

During a speech on 21 July 2004, Geoff Hoon announced a one-year delay to allow contractual and cost issues to be resolved. The building of the carriers was confirmed in December 2005. The building is being undertaken by four companies across seven shipyards, with final block integration and assembly at Rosyth:

BAE Systems Surface Ships – Govan (Lower Blocks 3 and 4), Scotstoun (aft island) and Portsmouth (Lower Block 2 and forward island)

Babcock Marine – Rosyth (Sponsons, Mast and Centre Blocks 5 and 6) and Appledore (Lower Block 1)

A&P Group – Hebburn (Centre Block 3)

Cammell Laird – Birkenhead (Centre Blocks 2 and 4)

In December 2007, eight diesel engines and electricity generators, four for each ship were ordered from Wärtsilä. On 4 March 2008, contracts for the supply of 80,000 tonnes of steel were awarded to Corus Group, with an estimated value of £65 million. Other contracts included £3 million for fibre optic cable, over £1 million for reverse osmosis equipment to provide over 500 tonnes of fresh water daily, and £4 million for aviation fuel systems. On 3 April 2008 a contract for the manufacture of aircraft lifts (worth £13m) was awarded to MacTaggart Scott of Loanhead, Scotland.

In mid May 2008, the Treasury announced that it would be making available further funds on top of the regular defence budget, reportedly allowing the construction of the carriers to begin. This was followed, on 20 May 2008, by the government giving the "green light" for construction of the Queen Elizabeth class, stating that it was ready to sign the contracts for full production once the creation of the planned shipbuilding joint venture between BAE Systems and the VT Group had taken place. This joint venture, BVT Surface Fleet, became operational on 1 July 2008. VT Group later sold its share to BAE Systems which renamed the unit BAE Systems Surface Ships. It will undertake approximately forty per cent of the project workload.

On 1 September 2008, the MoD announced a £51 million package of important equipment contracts; £34 million for the highly mechanised weapons handling system for the two ships, £8 million for supply of uptake and down-take systems for both ships, £5 million for air traffic control software, £3 million for supply of pumps and associated systems engineering, and £1 million for emergency diesel generators. On 6 October 2008, it was announced that contracts had been placed for "the carriers' Rolls-Royce gas turbines, generators, motors, power distribution equipment, platform management systems, propellers, shafts, steering gear, rudders and stabilisers".

The construction of the two carriers involves more than 10,000 people from 90 companies, 7,000 of them in the six shipyards building the sections of the ships.

The first steel cut for the project, in July 2009, signalled the start of construction of Lower Block 3 at BAE Systems Clyde, where production of Lower Block 4 started in January 2010. Meanwhile, construction of the bow Lower Block 1 was carried out at Appledore, North Devon, and was completed in March 2010. When the four lower blocks are completed they will be transported to Rosyth to be assembled.

On 25 January 2010, it was announced that the Cammell Laird shipyard has secured a £44 million contract to build the flight decks of the carriers. That same day, construction began in Portsmouth of Lower Block 2 for Queen Elizabeth. The structure will house machinery spaces, stores, switchboards and some of the ship's accommodation. The block will weigh around 6,000 tonnes and will stand over 18 metres (59 ft) tall, 70 metres (230 ft) long and 40 metres (130 ft) wide.

On 16 August 2011, the 8,000-tonne Lower Block 03 of Queen Elizabeth left BAE Systems Surface Ships' Govan shipyard in Glasgow on a large ocean-going barge. Travelling 600 miles (970 km) around the northern coast of Scotland, the block arrived at Rosyth on the evening of 20 August 2011.

Under the present plans Queen Elizabeth will enter service in 2016.

Construction on the second carrier, Prince of Wales, began on 26 May 2011 when then Defence Secretary Liam Fox cut the first steel. Although it had been intended to convert the Prince of Wales to a CATOBAR configuration, the government announced in May 2012 that the Short Take-Off and Landing F-35B variant jet would be purchased instead. The carrier will now be completed with a "ski-jump" ramp, in the STOVL configuration. As of November 2011, the carrier was expected to enter service in 2018.

US Military Aircraft

Attack Aircraft Variants

McDonnell Douglas F/A-18 Hornet

The McDonnell Douglas (now Boeing) F/A-18 Hornet is a twin-engine supersonic, all-weather carrier-capable multirole fighter jet, designed to dogfight and attack ground targets (F/A for Fighter/Attack). Designed by McDonnell Douglas and Northrop, the F/A-18 was derived from the latter's YF-17 in the 1970s for use by the United States Navy and Marine Corps. The Hornet is also used by the air forces of several other nations. It has been the aerial demonstration aircraft for the U.S. Navy's Flight Demonstration Squadron, the Blue Angels, since 1986.

The F/A-18 has a top speed of Mach 1.8. It can carry a wide variety of bombs and missiles, including air-to-air and air-to-ground, supplemented by the 20 mm M61 Vulcan cannon. It is powered by two General Electric F404 turbofan engines, which give the aircraft a high thrust-to-weight ratio. The F/A-18 has excellent aerodynamic characteristics, primarily attributed to its leading edge extensions (LEX). The fighter's primary missions are fighter escort, fleet air defense, Suppression of Enemy Air Defenses (SEAD), air interdiction, close air support and aerial reconnaissance. Its versatility and reliability have proven it to be a valuable carrier asset, though it has been criticized for its lack of range and payload compared to its earlier contemporaries, such as the Grumman F-14 Tomcat in the fighter and strike fighter role, and the Grumman A-6 Intruder and LTV A-7 Corsair II in the attack role.

The F/A-18 Hornet provided the baseline design for the Boeing F/A-18E/F Super Hornet, a larger, evolutionary redesign of the F/A-18. Compared to the Hornet, the Super Hornet is larger, heavier and has improved range and payload. The F/A-18E/F was originally proposed as an alternative to an all-new aircraft to replace existing dedicated attack aircraft such as the A-6. The larger variant was also directed to replace the aging F-14 Tomcat, thus serving a complementary role with Hornets in the U.S. Navy, and serving a wider range of roles including refueling tanker. The Boeing EA-18G Growler electronic jamming platform was also developed from the F/A-18E/F Super Hornet.

The F/A-18 is a twin engine, mid-wing, multi-mission tactical aircraft. It is highly maneuverable, owing to its good thrust to weight ratio, digital fly-by-wire control system, and leading edge extensions (LEX). The LEX allow the Hornet to remain controllable at high angles of attack. The wing is a trapezoidal shape with 20-degree sweepback on the leading edge and a straight trailing edge. The wing has full-span leading edge flaps and the trailing edge has single-slotted flaps and ailerons over the entire span.

Canted vertical stabilizers are another distinguishing design element, one among several other such elements that enable the Hornet's excellent high angle-of-attack ability include oversized horizontal stabilators, oversized trailing edge flaps that operate as flaperons, large full-length leading edge slats, and flight control computer programming that multiplies the movement of each control surface at low speeds and moves the vertical rudders inboard instead of simply left and right. The Hornet's normally high angle-of-attack performance envelope was put to rigorous testing and enhanced in the NASA F-18 High Alpha Research Vehicle (HARV). NASA used the F-18 HARV to demonstrate flight handling characteristics at high angle-of-attack (alpha) of 65–70 degrees using thrust vectoring vanes. F/A-18 stabilators were also used as canards on NASA's F-15S/MTD.

The Hornet was among the first aircraft to heavily use multi-function displays, which at the switch of a button allow a pilot to perform either fighter or attack roles or both. This "force multiplier" ability gives the operational commander more flexibility to employ tactical aircraft in a fast-changing battle scenario. It was the first Navy aircraft to incorporate a digital multiplex avionics bus, enabling easy upgrades.

The Hornet is also notable for having been designed to reduce maintenance, and as a result has required far less downtime than its heavier counterparts, the F-14 Tomcat and the A-6 Intruder. Its mean time between failure is three times greater than any other Navy strike aircraft, and requires half the maintenance time. Its General Electric F404 engines were also innovative in that they were designed with operability, reliability and maintainability first. The engine, while unexceptional in rated performance, demonstrates exceptional robustness under various conditions and is resistant to stall and flameout. The F404 engine connects to the airframe at only 10 points and can be replaced without special equipment; a four person team can remove the engine within 20 minutes.

The engine air inlets of the Hornet, like that of the F-16, are of a simpler "fixed" design, while those of the F-4, F-14, and F-15 have variable geometry or variable ramp air inlets. This is a speed limiting factor in the Hornet design. Instead, the Hornet uses bleed air vents on the inboard surface of the engine air intake ducts to slow and reduce the amount of air reaching the engine. While not as effective as variable geometry, the bleed air technique functions well enough to achieve near Mach 2 speeds, which is within the designed mission requirements.

A 1989 USMC study found that single seat fighters were well suited to air to air combat missions while dual seat fighters were favored for complex strike missions against heavy air and ground defenses in adverse weather. The question being not so much as to whether a second pair of eyes would be useful, but as to having the second crewman sit in the same fighter or in a second fighter. Single-seat fighters that lacked wingmen were shown to be especially vulnerable.

F/A-18 Hornet in transonic flight exhibiting Prandtl-Glauert condensation

McDonnell Douglas rolled out the first F/A-18A on 13 September 1978, in blue-on-white colors marked with "Navy" on the left and "Marines" on the right. Its first flight was on 18 November. In a break with tradition, the Navy pioneered the "principal site concept" with the F/A-18, where almost all testing was done at Naval Air Station Patuxent River, instead of near the site of manufacture, and using Navy and Marine Corps test pilots instead of civilians early in development. In March 1979, Lt. Cdr. John Padgett became the first Navy pilot to fly the F/A-18.

Following trials and operational testing by VX-4 and VX-5, Hornets began to fill the Fleet Replacement Squadrons (FRS) VFA-125, VFA-106, and VMFAT-101, where pilots are introduced to the F/A-18. The Hornet entered operational service with Marine Corps squadron VMFA-314 at MCAS El Toro on 7 January 1983, and with Navy squadron VFA-113 in March 1983, replacing F-4s and A-7Es, respectively.

The US Navy's Blue Angels Flight Demonstration Squadron switched to the F/A-18 Hornet in 1986, when it replaced the A-4 Skyhawk. The Blue Angels perform in F/A-18A and B models at air shows and other special events across the US and worldwide. Blue Angels pilots must have 1,350 hours and an aircraft carrier certification.

The two-seat B model is typically used to give rides to VIPs, but can also fill in for other aircraft in the squadron in a normal show, if the need arises.

Lockheed AC-130

The Lockheed AC-130 gunship is a heavily-armed ground-attack aircraft variant of the C-130 Hercules transport plane. The basic airframe is manufactured by Lockheed, while Boeing is responsible for the conversion into a gunship and for aircraft support. The AC-130A Gunship II superseded the AC-47 Gunship I during the Vietnam War.

The gunship's sole user is the United States Air Force, which uses AC-130H Spectre and AC-130U Spooky variants for close air support, air interdiction and force protection. Close air support roles include supporting ground troops, escorting convoys, and flying urban operations. Air interdiction missions are conducted against planned targets and targets of opportunity. Force protection missions include defending air bases and other facilities. AC-130Us are based at Hurlburt Field, Florida and AC-130Hs are based at Cannon AFB, New Mexico; though both deploy to bases worldwide in support of operations. The gunship squadrons are part of the Air Force Special Operations Command (AFSOC), a component of the United States Special Operations Command (SOCOM).

Most of the weaponry aboard is mounted to fire out from the left or port side of the aircraft. During an attack, the gunship performs a pylon turn over the target area (flying in a large circle around a fixed point on the ground, the fixed point being the target). This allows it to maintain fire at a target far longer than a conventional attack aircraft. The AC-130H "Spectre" can be armed with two 20 mm M61 Vulcan cannons, one Bofors 40mm autocannon, and one 105 mm M102 cannon. The upgraded AC-130U "Spooky" has a single 25 mm GAU-12 Equalizer in place of the Spectre's twin 20 mm cannons, as well as an improved fire control system and increased capacity for ammunition. Power is provided by four Allison T56-A-15 turboprops (standard for a C-130 Hercules). New AC-130J gunships based on MC-130J Combat Shadow II special operations tankers are planned.

Bell AH-1 Cobra

The Bell AH-1 SuperCobra is a twin-engine attack helicopter based on the US Army's AH-1 Cobra. The twin Cobra family includes the *AH-1J SeaCobra*, the AH-1T Improved SeaCobra, and the AH-1W SuperCobra. The AH-1W is the backbone of the United States Marine Corps's attack helicopter fleet, but will be replaced in service by the Bell AH-1Z Viper upgrade.

Boeing AH-64 Apache

The Boeing AH-64 Apache is a four-blade, twin-engine attack helicopter with a tailwheel-type landing gear arrangement, and a tandem cockpit for a two-man crew. The Apache was developed as Model 77 by Hughes Helicopters for the United States Army's Advanced Attack Helicopter program to replace the AH-1 Cobra, and was first flown on 30 September 1975. The AH-64 was introduced to US Army service in April 1986.

The AH-64 Apache features a nose-mounted sensor suite for target acquisition and night vision systems. It is armed with a 30-millimeter (1.2 in) M230 Chain Gun carried between the main landing gear, under the aircraft's forward fuselage. It has four hardpoints mounted on stub-wing pylons, typically carrying a mixture of AGM-114 Hellfire missiles and Hydra 70 rocket pods. The AH-64 has a large amount of systems redundancy to improve combat survivability.

The U.S. Army selected the AH-64, by Hughes Helicopters, over the Bell YAH-63 in 1976, and later approved full production in 1982. McDonnell Douglas continued production and development after purchasing Hughes Helicopters from Summa Corporation in 1984. The first production AH-64D Apache Longbow, an upgraded version of the original Apache, was delivered to the Army in March 1997. Production has been continued by Boeing Defense, Space & Security; over one thousand AH-64s have been produced to date.

The U.S. Army is the primary operator of the AH-64; it has also become the primary attack helicopter of multiple nations, including Greece, Japan, Israel, the Netherlands and Singapore; as well as being produced under license in the United Kingdom as the AgustaWestland Apache. U.S. AH-64s have served in conflicts in Panama, the Persian Gulf, Kosovo, Afghanistan, and Iraq. Israel has made active use of the Apache in its military conflicts in Lebanon and the Gaza Strip, while two coalition allies have deployed their AH-64s in Afghanistan and Iraq.

Northrop T-38 Talon

The Northrop T-38 Talon is a twin-engine supersonic jet trainer. It was the world's first supersonic trainer and is also the most produced. The T-38 remains in service as of 2012 in air forces throughout the world.

The United States Air Force (USAF) is the largest operator of the T-38. In addition to training USAF pilots, the T-38 is used by NASA. The US Naval Test Pilot School is the principal US Navy operator (other T-38s were previously used as USN aggressor aircraft until replaced by the similar Northrop F-5 Tiger II). Pilots of other NATO nations fly the T-38 in joint training programs with USAF pilots.

The basic airframe was used for the light combat aircraft F-5 Freedom Fighter family. In the 1950s Northrop began studying lightweight and more affordable fighter designs. The company began with its single-engine Northrop N-102 Fang concept. The N-102 was facing weight and cost growth, so the project was canceled and the company N-156 project was begun.

Although the USAF had no need for a small fighter at the time, it became interested in the trainer as a replacement for the T-33 Shooting Star it was then using in that role. The first of three prototypes (designated YT-38) flew on 10 March 1959.

The type was quickly adopted and the first production examples were delivered in 1961, officially entering service on 17 March that year, complementing the T-37 primary jet trainer. When production ended in 1972, 1,187 T-38s had been built. Since its introduction, it is estimated that some 50,000 military pilots have trained on this aircraft. The USAF remains one of the few armed flying forces using dedicated supersonic final trainers, as most, such as the US Navy, use high subsonic trainers.

The T-38 is of conventional configuration, with a small, low, long-chord wing, a single vertical stabilizer, and tricycle undercarriage. The aircraft seats a student pilot and instructor in tandem, and has intakes for its two turbojet engines at the wing roots. Its nimble performance has earned it the nickname white rocket. In 1962 the T-38 set absolute time-to-climb records for 3000, 6000, 9000 and 12000 meters, beating the records for those altitudes set by the F-104 in December 1958. (The F-4 beat the T-38's records less than a month later.)

The F-5B and F (which also derive from the N-156) can be distinguished from the T-38 by the wings; the wing of the T-38 meets the fuselage straight and ends square, while the F-5 has leading edge extensions near the wing roots and wingtip launch rails for air to air missiles. Under the paint the T-38 wing is constructed of honeycomb material while the wing of the F-5 family uses conventional skin over underlying support structure.

Most T-38s built were of the T-38A variant, but the USAF also had a small number of aircraft that had been converted for weapons training. These aircraft (designated AT-38B) had been fitted with a gunsight and could carry a gunpod, rockets, or bombs on a centerline pylon. In 2003, 562 T-38s were still operational with the USAF and are currently undergoing structural and avionics programs (T-38C) to extend their service life to 2020. Improvements include the addition of a HUD, GPS, INS (Inertial Navigation System), and TCAS as well as PMP (a propulsion modification to improve low-altitude engine thrust). Many USAF variants (T-38A and AT-38B) are being converted to the T-38C standard.

The fighter version of the N-156 was eventually selected for the US Military Assistance Program and produced as the F-5 Freedom Fighter. Many of these have since reverted to a weapons training role as various air forces have introduced newer types into service. The F-5G was an advanced single engine variant later renamed the F-20 Tigershark.

McDonnell Douglas AV-8B Harrier II

The McDonnell Douglas (now Boeing) AV-8B Harrier II is a second-generation vertical/short takeoff and landing (V/STOL) ground-attack aircraft. An Anglo-American development of the British Hawker Siddeley Harrier, the Harrier II is the final member of the Harrier family that started with the Hawker Siddeley P.1127 in the early 1960s. The AV-8B is primarily used for light attack or multi-role missions, and is typically operated from small aircraft carriers, large amphibious assault ships and simple forward operating bases. The AV-8B is used by the United States Marine Corps (USMC), Spanish Navy and Italian Navy. A variant of the AV-8B, the British Aerospace Harrier II was developed for the British military. The TAV-8B is a dedicated two-seat trainer version. The Harrier II and other models of the Harrier family have been called "Jump Jets".

The AV-8B was extensively redesigned by McDonnell Douglas from the earlier AV-8A/C Harrier. It has a new wing, an elevated cockpit, a redesigned fuselage, and other structural and aerodynamic refinements. The number of hardpoints was increased from five to seven. Later upgrades, which resulted in the AV-8B(NA) and AV-8B Harrier II Plus, added radar and night-attack capabilities. British Aerospace joined the improved Harrier project as partner in 1981. Since corporate mergers in the 1990s, Boeing and BAE Systems have jointly supported the program.

AV-8Bs have participated in numerous conflicts, providing close air support for ground troops and performing armed reconnaissance, proving themselves versatile assets. US Army General Norman Schwarzkopf named the USMC Harrier as one of the seven most important weapons of the Gulf War. The aircraft took part in combat during the Iraq War beginning in 2003. The Harrier II has served in Operation Enduring Freedom in Afghanistan since 2001, and was used in Operation Odyssey Dawn in Libya in 2011. Italian and Spanish Harrier IIs participated in overseas conflicts, in conjunction with NATO coalitions. American and Italian AV-8Bs are expected to be replaced by the Lockheed Martin F-35B Lightning II.

McDonnell Douglas AV-16+ Harrier

The AV-16 programme was launched in the early 1970s to develop an aircraft that was double the AV-8 - specifically, range and payload. The programme defined two aircraft - the subsonic Hawker P1184, and the supersonic Hawker P1185. The US effort focused on the airframe, and the UK effort on development of the pegasus engine. After a couple of years, the UK government pulled out as they did not have a firm commitment from the RAF - who were not convinced about a VSTOL solution to replace the Harrier and Jaguar. McDonnell Douglas pressed ahead and developed the AV-8B without UK support. Five years later, still without a solution to the Harrier/Jaguar problem, the UK returned to St Louis cap in hand to develop the Harrier GR5 from MDD's AV-8B...thus proceeded the US enthusiasm and UK apathy to the Harrier!

Bomber Aircraft Variants

Rockwell B-1 Lancer

The Rockwell (now part of Boeing) B-1 Lancer is a four-engine variable-sweep wing strategic bomber used by the United States Air Force (USAF). First envisioned in the 1960s as a supersonic bomber with sufficient range and payload to replace the Boeing B-52 Stratofortress, it developed primarily into a low-level penetrator with long range and supersonic speed capability at high altitude.

Designed by Rockwell International, the bomber's development was delayed multiple times over its history, as the theory of strategic balance changed from flexible response to mutually assured destruction and back again. This change in stance repeatedly demanded then ignored the need for manned bombers. The initial B-1A version was developed in the early 1970s, but its production was canceled, and only four prototypes were built. The need for a new

85

platform once again surfaced in the early 1980s, and the aircraft resurfaced as the B-1B version with the focus on low-level penetration bombing. However by this point development of stealth technology was promising an aircraft of dramatically improved capability. Production went ahead as the B-1B would be operational before these "Advanced Technology Bomber" concepts, during a period when the B-52 would be increasingly vulnerable. It entered service in 1986 with the USAF Strategic Air Command as a nuclear bomber.

In the 1990s, the B-1B was converted to conventional bombing use. It first served in combat during Operation Desert Fox in 1998 and again during the NATO action in Kosovo the following year. The B-1B has supported U.S. and NATO military forces in Afghanistan and Iraq. The Lancer is the supersonic component of the USAF's long-range bomber force, along with the subsonic B-52 and Northrop Grumman B-2 Spirit. The bomber is commonly called the "Bone" (originally from "B-One"). With the retirement of the General Dynamics/Grumman EF-111A Raven in 1998 and the Grumman F-14 Tomcat in 2006, the B-1B is the U.S. military's only active variable-sweep wing aircraft. The B-1B is expected to continue to serve into the 2020s, when it is to be supplemented by the Next Generation Bomber.

The B-1A's engine was modified slightly to produce the GE F101-102 for the B-1B, with an emphasis on durability, and increased efficiency. The core of this engine has since been re-used in several other engine designs, including the GE F110 which has seen use in the F-14 Tomcat, F-15K/SG variants and most recent versions of the General Dynamics F-16 Fighting Falcon. It is also the basis for the non-afterburning GE F118 used in the B-2 Spirit and the U-2S. However its greatest success was forming the core of the extremely popular CFM56 civil engine, which can be found on some versions of practically every small-to-medium sized airliner. The nose gear cover door has controls for the auxiliary power units (APUs), which allow for quick starts of the APUs upon order to scramble.

The B-1's main computer is the IBM AP-101, which is also used on the Space Shuttle orbiter and the B-52 bomber. The computer is programmed with the JOVIAL programming language. The Lancer's offensive avionics include the Westinghouse (now Northrop Grumman) AN/APQ-164 forward-looking offensive passive electronically scanned array radar set with electronic beam steering (and a fixed antenna pointed downward for reduced radar observability), synthetic aperture radar, ground moving target indicator (MTI), and terrain-following radar modes, Doppler navigation, radar altimeter, and an inertial navigation suite. The B-1B Block D upgrade added a Global Positioning System (GPS) receiver beginning in 1995.

A B-1B makes a high-speed, transonic pass at the Pensacola Beach air show, 2003

Northrop Grumman B-2 Spirit

The Northrop Grumman B-2 Spirit (also known as the Stealth Bomber) is an American strategic bomber, featuring low observable stealth technology designed for penetrating dense anti-aircraft defenses; it is able to deploy both conventional and nuclear weapons. The bomber has a crew of two and can drop up to eighty 500 lb (230 kg)-class JDAM GPS-guided bombs, or sixteen 2,400 lb (1,100 kg) B83 nuclear bombs. The B-2 is the only aircraft that can carry large air to surface standoff weapons in a stealth configuration.

Development originally started under the "Advanced Technology Bomber" (ATB) project during the Carter administration, and its performance was one of the reasons for his cancellation of the B-1 Lancer. ATB continued during the Reagan administration, but worries about delays in its introduction led to the reinstatement of the B-1 program as well. Program costs rose throughout development. Designed and manufactured by Northrop Grumman with assistance from Boeing, the cost of each aircraft averaged US$737 million (in 1997 dollars). Total procurement costs averaged $929 million per aircraft, which includes spare parts, equipment, retrofitting, and software support. The total program cost including development, engineering and testing, averaged $2.1 billion per aircraft in 1997.

Because of its considerable capital and operational costs, the project was controversial in the U.S. Congress and among the Joint Chiefs of Staff. The winding-down of the Cold War in the latter portion of the 1980s dramatically reduced the need for the aircraft, which was designed with the intention of penetrating Soviet airspace and attacking high-value targets. During the late 1980s and 1990s, Congress slashed initial plans to purchase 132 bombers to 21. In 2008, a B-2 was destroyed in a crash shortly after takeoff, and the crew ejected safely. A total of 20 B-2s remain in service with the United States Air Force.

Though originally designed primarily as a nuclear bomber, the B-2 was first used in combat to drop conventional bombs on Serbia during the Kosovo War in 1999, and saw continued use during the wars in Iraq and Afghanistan. B-2s were also used during the 2011 Libyan civil war.

Boeing B-52 Stratofortress

The Boeing B-52 Stratofortress is a long-range, subsonic, jet-powered strategic bomber. The B-52 was designed and built by Boeing, who have continued to provide support and upgrades. It has been operated by the United States Air Force (USAF) since the 1950s. The bomber carries up to 70,000 pounds (32,000 kg) of weapons.

Beginning with the successful contract bid in June 1946, the B-52 design evolved from a straight-wing aircraft powered by six turboprop engines to the final prototype YB-52 with eight turbojet engines and swept wings. The B-52 took its maiden flight in April 1952. Built to carry nuclear weapons for Cold War-era deterrence missions, the B-52 Stratofortress replaced the Convair B-36. Although a veteran of a number of wars, the Stratofortress has dropped only conventional munitions in combat.

The B-52 has been in active service with the USAF since 1955. The bombers flew under the Strategic Air Command (SAC) until it was disestablished in 1992 and its aircraft absorbed into the Air Combat Command (ACC); in 2010 all B-52 Stratofortresses were transferred from the ACC to the new Air Force Global Strike Command (AFGSC). Superior performance at high subsonic speeds and relatively low operating costs have kept the B-52 in service despite the advent of later aircraft, including the Mach 3 North American XB-70 Valkyrie, the variable-geometry Rockwell B-1B Lancer, and the stealthy Northrop Grumman B-2 Spirit. The B-52 marked its 50th anniversary of continuous service with its original operator in 2005 and after being upgraded between 2013 and 2015 it will serve into the 2040s.

Cargo Aircraft Variants

Lockheed C-5 Galaxy

The Lockheed C-5 Galaxy is a large military transport aircraft built by Lockheed. It provides the United States Air Force (USAF) with a heavy intercontinental-range strategic airlift capability, one that can carry outsize and oversize cargos, including all air-certifiable cargo. The Galaxy has many similarities to its smaller C-141 Starlifter predecessor, and the later C-17 Globemaster. The C-5 is among the largest military aircraft in the world.

The C-5 Galaxy had a complicated development; significant cost overruns were experienced and Lockheed suffered significant financial difficulties. Shortly after entering service, fractures in the wings of many aircraft were discovered and the C-5 fleet were restricted in capability until corrective work was conducted. The C-5M Super Galaxy is an upgraded version with new engines and modernized avionics designed to extend its service life beyond 2040.

The C-5 Galaxy has been operated by USAF since 1969. In that time, it has been used to support US military operations in all major contingencies including Vietnam, Iraq, Yugoslavia and Afghanistan; as well as in support of allies, such as Israel during the Yom Kippur War and NATO operations in the Gulf War. The C-5 has also been used to distribute humanitarian aid and disaster relief, and in support of the US Space Shuttle program run by NASA.

C-7A   DEHAVILLAND CARIBOU

The de Havilland Canada DHC-4 Caribou (designated by the United States military as the CV-2 and later C-7 Caribou) is a Canadian-designed and produced specialized cargo aircraft with short takeoff and landing (STOL) capability. The Caribou was first flown in 1958 and although mainly retired from military operations, is still in use in small numbers as a rugged "bush" aircraft.

The de Havilland Canada company's third STOL design was a big step up in size compared to its earlier DHC Beaver and DHC Otter, and was the first DHC design powered by two engines. The Caribou, however, was similar in concept in that it was designed as a rugged STOL utility. The Caribou was primarily a military tactical transport that in commercial service found itself a small niche in cargo hauling. The United States Army ordered 173 in 1959 and took delivery in 1961 under the designation AC-1, which was changed to CV-2 Caribou in 1962.

Boeing C-17 Globemaster III

The Boeing C-17 Globemaster III is a large military transport aircraft. It was developed for the United States Air Force (USAF) from the 1980s to the early 1990s by McDonnell Douglas; the company later merged with Boeing. The C-17 is used for rapid strategic airlift of troops and cargo to main operating bases or forward operating bases throughout the world. It can also perform tactical airlift, medical evacuation and airdrop missions. The C-17 carries the name of two previous, but unrelated piston-engine, U.S. military cargo aircraft, the Douglas C-74 Globemaster and the Douglas C-124 Globemaster II.

In addition to the U.S. Air Force, the C-17 is operated by the United Kingdom, Australia, Canada, Qatar, United Arab Emirates and NATO Heavy Airlift Wing. Additionally, India has ordered C-17s.

The C-17 is 174 feet (53 m) long and has a wingspan of about 170 feet (52 m). It can airlift cargo fairly close to a battle area. The size and weight of U.S. mechanized firepower and equipment have grown in recent decades from increased air mobility requirements, particularly for large or heavy non-palletized outsize cargo.

The C-17 is powered by four Pratt & Whitney F117-PW-100 turbofan engines, which are based on the commercial Pratt and Whitney PW2040 used on the Boeing 757. Each engine is fully reversible and rated at 40,400 lbf (180 kN) of thrust. The thrust reversers direct engine exhaust air upwards and forward, reducing the chances of foreign object damage by ingestion of runway debris, and providing enough reverse thrust to back the aircraft up on the ground while taxiing. The thrust reversers can also be used in flight at idle-reverse for added drag in maximum-rate descents.

C-21A GATES LEARJET

The Learjet Model 35 and Model 36 are a series of American multi-role business jets and military transport aircraft. When used by the United States Air Force they carry the designation C-21A.

The aircraft are powered by two Garrett TFE731-2 turbofan engines. Its cabin can be arranged for 6-8 passengers.

The Model 36 has a shortened passenger area in the fuselage, in order to provide more space in the aft fuselage for fuel tanks. It is designed for longer-range mission capability.

The engines are mounted in nacelles on the sides of the aft fuselage. The wings are equipped with single-slotted flaps. The wingtip fuel tanks distinguish the design from other aircraft having similar functions.

The concept which became the LJ35 began as the Learjet 25BGF (with GF referring to "Garrett Fan"), a Learjet 25 with a then-new TFE731 turbofan engine mounted on the left side in place of the 25's General Electric CJ610 turbojet engine.

This testbed aircraft first flew in May, 1971. As a result of the increased power and reduced noise of the new engine, Learjet further improved the design, and instead of being simply a variant of the 25, it became its own model, the 35.

C-22B BOEING

The Boeing 727 is a mid-size narrow-body three-engine jet airliner built by Boeing Commercial Airplanes. It can carry 149 to 189 passengers and later models can fly up to 2,400 to 2,700 nautical miles (4,400 to 5,000 km) nonstop. Intended for short and medium-length flights, the 727 can use fairly short runways at smaller airports.

It has three Pratt & Whitney JT8D engines below the T-tail, one on each side of the fuselage with a center engine that connects through an S-duct to an inlet at the base of the fin. The 727 followed the 707 quad-jet airliner with which it shares its upper fuselage cross-section and cockpit design.

The 727-100 first flew in February 1963 and entered service with Eastern Air Lines in February 1964; the stretched 727-200 flew in July 1967 and entered service with Northeast Airlines that December.

The 727 became a mainstay of airlines' domestic route networks and was also used on short- and medium-range international routes.

Passenger, freighter, and convertible versions of the 727 were built.

The 727 was heavily produced into the 1970s; the last 727 was completed in 1984. In July 2011, 23 727-100s and 227 727-200s were in airline service. Airport noise regulations have led to 727s being equipped with hush kits.

C-23B SHORT BROTHERS LTD

The Short C-23 Sherpa is a small military transport aircraft built by Short Brothers. The C-23A and C-23B variants are variants of the Short 330 and the C-23B+ and C-23C are variants of the Short 360.

C-47T   DOUGLAS ACFT / BASKET

The Douglas C-47 Skytrain or Dakota is a military transport aircraft that was developed from the Douglas DC-3 airliner. It was used extensively by the Allies during World War II and remained in front line operations through the 1950s with a few remaining in operation.

Boeing Vertol CH-46 Sea Knight

The Boeing Vertol CH-46 Sea Knight is a medium-lift tandem rotor transport helicopter. It is used by the United States Marine Corps (USMC) to provide all-weather, day-or-night assault transport of combat troops, supplies and equipment. Additional tasks include combat support, search and rescue (SAR), support for forward refueling and rearming points, CASEVAC and Tactical Recovery of Aircraft and Personnel (TRAP). Canada also operated the Sea Knight, designated as CH-113, and operated them in the SAR role until 2004. Other export customers include Japan, Sweden, and Saudi Arabia. The commercial version is the BV 107-II, commonly referred to simply as the "Vertol".

The CH-46 has tandem contrarotating rotors powered by two GE T58 turboshaft engines. The engines are mounted on each side of the rear rotor pedestal with a driveshaft to the forward rotor. The engines are coupled so either could power both rotors in an emergency. The rotors feature three blades and can be folded for on-ship operations. The CH-46 has fixed tricycle landing gear, with twin wheels on all three units. The gear configuration causes a nose-up stance to facilitate cargo loading and unloading. The main gear are fitted in rear sponsons that also contain fuel tanks with a total capacity of 350 US gallons (1,438 L).

The CH-46 has a cargo bay with a rear loading ramp that could be removed or left open in flight for extended cargo or for parachute drops. An internal winch is mounted in the forward cabin and can be used to pull external cargo on pallets into the aircraft via the ramp and rollers. A belly sling hook (cargo hook) which is usually rated at 10,000 lb (4,500 kg). could be attached for carrying external cargo. Although the hook is rated at 10,000 lb (4,500 kg)., the limited power produced by the engines preclude the lifting of such weight. It usually has a crew of three, but can accommodate a larger crew depending on mission specifics. For example, a Search and Rescue variant will usually carry a crew of five (Pilot, Co-Pilot, Crew Chief, Swimmer, and Medic) to facilitate all aspects of such a mission. A pintle-mounted 0.50 in (12.7 mm) Browning machine gun is mounted on each side of the helicopter for self-defense. Service in southeast Asia resulted in the addition of armor with the guns.

Boeing CH-47 Chinook

The Boeing CH-47 Chinook is an American twin-engine, tandem rotor heavy-lift helicopter. With a top speed of 170 knots (196 mph, 315 km/h) it is faster than contemporary utility and attack helicopters of the 1960s.

Alongside the C-130 Hercules and the UH-1 Iroquois, the CH-47 is one of the few aircraft of that era that is still in production and front line service, with over 1,179 built to date. Its primary roles include troop movement, artillery emplacement and battlefield resupply. It has a wide loading ramp at the rear of the fuselage and three external-cargo hooks.

The Chinook was designed and initially produced by Boeing Vertol in the early 1960s. The helicopter is now produced by Boeing Rotorcraft Systems. Chinooks have been sold to 16 nations with the US Army and the Royal Air Force (see Boeing Chinook (UK variants)) being the largest users. The CH-47 is among the heaviest lifting Western helicopters.

Sikorsky CH-53E Super Stallion

The Sikorsky CH-53E Super Stallion is the largest and heaviest helicopter in the United States military. As the Sikorsky S-80 it was developed from the CH-53 Sea Stallion, mainly by adding a third engine, a seventh blade to the main rotor and canting the tail rotor 20 degrees.

It was built by Sikorsky Aircraft for the United States Marine Corps. The less common MH-53E Sea Dragon fills the United States Navy's need for long range mine sweeping or Airborne Mine Countermeasures (AMCM) missions, and perform heavy-lift duties for the Navy. Under development is the CH-53K, which will be equipped with new engines, new composite rotor blades, and a wider cabin.

Bell Boeing V-22 Osprey

The Bell Boeing V-22 Osprey is an American multi-mission, military, tiltrotor aircraft with both a vertical takeoff and landing (VTOL), and short takeoff and landing (STOL) capability. It is designed to combine the functionality of a conventional helicopter with the long-range, high-speed cruise performance of a turboprop aircraft.

The V-22 originated from the United States Department of Defense Joint-service Vertical take-off/landing Experimental (JVX) aircraft program started in 1981. The team of Bell Helicopter and Boeing Helicopters was awarded a development contract in 1983 for the tiltrotor aircraft. The Bell Boeing team jointly produce the aircraft. The V-22 first flew in 1989, and began flight testing and design alterations; the complexity and difficulties of being the first tiltrotor intended for military service in the world led to many years of development.

The United States Marine Corps began crew training for the Osprey in 2000, and fielded it in 2007; it is supplementing and will eventually replace their CH-46 Sea Knights. The Osprey's other operator, the U.S. Air Force, fielded their version of the tiltrotor in 2009. Since entering service with the U.S. Marine Corps and Air Force, the Osprey has been deployed in both combat and rescue operations over Iraq, Afghanistan and Libya.

Fighter Aircraft Variants

McDonnell Douglas F-4 Phantom II

The McDonnell Douglas F-4 Phantom II is a tandem, two-seat, twin-engine, all-weather, long-range supersonic jet interceptor fighter/fighter-bomber originally developed for the United States Navy by McDonnell Aircraft. It first entered service in 1960 with the U.S. Navy. Proving highly adaptable, it was also adopted by the U.S. Marine Corps and the U.S. Air Force, and by the mid-1960s had become a major part of their respective air wings.

The Phantom is a large fighter with a top speed of over Mach 2.2. It can carry over 18,000 pounds (8,400 kg) of weapons on nine external hardpoints, including air-to-air missiles, air-to-ground missiles, and various bombs. The F-4, like other interceptors of its time, was designed without an internal cannon. Later models incorporated a M61 Vulcan rotary cannon. Beginning in 1959 it set 15 world records for in-flight performance, including an absolute speed record, and an absolute altitude record.

The F-4 was used extensively during the Vietnam War, serving as the principal air superiority fighter for both the Navy and Air Force, as well as being important in the ground-attack and reconnaissance roles by the close of U.S. involvement in the war. The Phantom has the distinction of being the last U.S. fighter flown to attain ace status in the 20th century. During the Vietnam War the USAF had one pilot and two weapon systems officers (WSOs), and the US Navy one pilot and one radar intercept officer (RIO), achieve five aerial kills against other enemy fighter aircraft and become aces in air-to-air combat. The F-4 continued to form a major part of U.S. military air power throughout the 1970s and 1980s, being gradually replaced by more modern aircraft such as the F-15 Eagle and F-16 in the U.S. Air Force; the Grumman F-14 Tomcat and F/A-18 Hornet in the U.S. Navy; and the F/A-18 in the U.S. Marine Corps.

The F-4 Phantom II remained in use by the U.S. in the reconnaissance and Wild Weasel (suppression of enemy air defenses) roles in the 1991 Gulf War, finally leaving service in 1996. It was also the only aircraft used by both U.S. flight demonstration teams: the USAF Thunderbirds (F-4E) and the US Navy Blue Angels (F-4J). The F-4 was also operated by the armed forces of 11 other nations. Israeli Phantoms saw extensive combat in several Arab–Israeli conflicts, while Iran used its large fleet of Phantoms in the Iran–Iraq War.

Phantoms remain in front line service with seven countries, and in use as an unmanned target in the U.S. Air Force. Phantom production ran from 1958 to 1981, with a total of 5,195 built, making it the most numerous American supersonic military aircraft.

The F-4 Phantom is a tandem-seat fighter-bomber designed as a carrier-based interceptor to fill the U.S. Navy's fleet defense fighter role. Innovations in the F-4 included an advanced pulse-doppler radar and extensive use of titanium in its airframe.

Northrop F-5

The Northrop F-5A/B Freedom Fighter and the F-5E/F Tiger II are part of a family of widely-used light supersonic fighter aircraft, designed and built by Northrop. Hundreds remain in service in air forces around the world in the early 21st century, and the type has also been the basis for a number of other aircraft.

The F-5 started life as a privately-funded light fighter program by Northrop in the 1950s. The first-generation F-5A Freedom Fighter entered service in the 1960s. During the Cold War, over 800 were produced through 1972 for U.S. allies and Switzerland. The USAF had no need for a light fighter but specifed a requirement for a supersonic trainer, procuring about 1,200 of a derivative airframe for this purpose, the Northrop T-38 Talon.

The improved second-generation F-5E Tiger II was also primarily used by American Cold War allies and, in limited quantities, served in U.S. military aviation as a training and aggressor aircraft; Tiger II production amounted to 1,400 of all versions, with production ending in 1987. Many F-5s continuing in service into the 1990s and 2000s have undergone a wide variety of upgrade programs to keep pace with the changing combat environment.

The F-5 was also developed into a dedicated reconnaissance version, the RF-5 Tigereye. The F-5 also served as a starting point for a series of design studies which resulted in the twin-tailed Northrop YF-17 and the F/A-18 series of carrier-based fighters. The Northrop F-20 Tigershark was an advanced version of the F-5E that did not find a market. The F-5N/F variants remain in service with the United States Navy and United States Marine Corps as an adversary trainer.

In 1970, Northrop won the International Fighter Aircraft (IFA) competition to replace the F-5A, with better air-to-air performance against aircraft like the Soviet MiG-21. The resultant aircraft, initially known as F-5A-21, subsequently became the F-5E. It had more powerful (5,000 lbf) General Electric J85-21 engines, and had a lengthened and enlarged fuselage, accommodating more fuel. Its wings were fitted with enlarged leading edge extensions, giving an increased wing area and improved maneuverability. The aircraft's avionics were more sophisticated, crucially including a radar (initially the Emerson Electric AN/APQ-153) (the F-5A and B had no radar). It retained the gun armament of two M39 cannon, one on either side of the nose) of the F-5A. Various specific avionics fits could be accommodated at customer request, including an inertial navigation system, TACAN and ECM equipment.

Grumman F-14 Tomcat

The Grumman F-14 Tomcat is a supersonic, twin-engine, two-seat, variable-sweep wing fighter aircraft. The Tomcat was developed for the United States Navy's Naval Fighter Experimental (VFX) program following the collapse of the F-111B project. The F-14 was the first of the American teen-series fighters which were designed incorporating the experience of air combat against MiG fighters during the Vietnam War.

The F-14 first flew in December 1970 and made its first deployment in 1974 with the U.S. Navy aboard USS Enterprise (CVN-65), replacing the McDonnell Douglas F-4 Phantom II. The F-14 served as the U.S. Navy's primary maritime air superiority fighter, fleet defense interceptor and tactical reconnaissance platform. In the 1990s, it added the Low Altitude Navigation and Targeting Infrared for Night (LANTIRN) pod system and began performing precision ground-attack missions. The Tomcat was retired from the active U.S. Navy fleet on 22 September 2006, having been supplanted by the Boeing F/A-

18E/F Super Hornet. As of 2012, the F-14 was only in service with the Islamic Republic of Iran Air Force, having been exported to Iran in 1976, when the U.S. had amicable diplomatic relations with the then government of Shah Mohammad Reza Pahlavi.

McDonnell Douglas F-15 Eagle

The McDonnell Douglas (now Boeing) F-15 Eagle is a twin-engine, all-weather tactical fighter designed by McDonnell Douglas to gain and maintain air superiority in aerial combat. It is considered among the most successful modern fighters, with over 100 aerial combat victories with no losses in dogfights. Following reviews of proposals, the United States Air Force selected McDonnell Douglas' design in 1967 to meet the service's need for a dedicated air superiority fighter. The Eagle first flew in July 1972, and entered service in 1976.

Since the 1970s, the Eagle has also been exported to Israel, Japan, and Saudi Arabia. Despite originally being envisioned as a pure air superiority aircraft, the design proved flexible enough that an all-weather strike derivative, the F-15E Strike Eagle, was later developed, and entered service in 1989. The F-15 Eagle is expected to be in service with the U.S. Air Force past 2025.

General Dynamics F-16 Fighting Falcon

The General Dynamics F-16 Fighting Falcon is a multirole jet fighter aircraft originally developed by General Dynamics for the United States Air Force (USAF). Designed as an air superiority day fighter, it evolved into a successful all-weather multirole aircraft. Over 4,500 aircraft have been built since production was approved in 1976. Although no longer being purchased by the U.S. Air Force, improved versions are still being built for export customers. In 1993,

General Dynamics sold its aircraft manufacturing business to the Lockheed Corporation, which in turn became part of Lockheed Martin after a 1995 merger with Martin Marietta.

The Fighting Falcon is a fighter with numerous innovations including a frameless bubble canopy for better visibility, side-mounted control stick to ease control while maneuvering, a seat reclined 30 degrees to reduce the effect of g-forces on the pilot, and the first use of a relaxed static stability/fly-by-wire flight control system that makes it a highly nimble aircraft. The F-16 has an internal M61 Vulcan cannon and has 11 hardpoints for mounting weapons and other mission equipment. The F-16's official name is "Fighting Falcon", but "Viper" is commonly used by its pilots, due to a perceived resemblance to a viper snake as well as the Battlestar Galactica Colonial Viper starfighter.

In addition to active duty U.S. Air Force, Air Force Reserve Command, and Air National Guard units, the aircraft is also used by the USAF aerial demonstration team, the U.S. Air Force Thunderbirds, and as an adversary/aggressor aircraft by the United States Navy. The F-16 has also been procured to serve in the air forces of 25 other nations.

Lockheed Martin F-22 Raptor

The Lockheed Martin/Boeing F-22 Raptor is a single-seat, twin-engine fifth-generation supermaneuverable fighter aircraft that uses stealth technology. It was designed primarily as an air superiority fighter, but has additional capabilities that include ground attack, electronic warfare, and signals intelligence roles. Lockheed Martin Aeronautics is the prime contractor and is responsible for the majority of the airframe, weapon systems and final assembly of the F-22. Program partner Boeing Defense, Space & Security provides the wings, aft fuselage, avionics integration, and training systems.

The aircraft was variously designated F-22 and F/A-22 during the years prior to formally entering USAF service in December 2005 as the F-22A. Despite a protracted and costly development period, the United States Air Force considers the F-22 a critical component of U.S. tactical air power, and claims that the aircraft is unmatched by any known or projected fighter, while Lockheed Martin claims that the Raptor's combination of stealth, speed, agility, precision and situational awareness, combined with air-to-air and air-to-ground combat capabilities, makes it the best overall fighter in the world today. Air Chief Marshal Angus Houston, former Chief of the Australian Defence Force, said in 2004 that the "F-22 will be the most outstanding fighter plane ever built."

The high cost of the aircraft, a lack of clear air-to-air combat missions because of delays in the Russian and Chinese fifth-generation fighter programs, a U.S. ban on Raptor exports, and the ongoing development of the planned cheaper and more versatile F-35 resulted in calls to end F-22 production. In April 2009 the US Department of Defense proposed to cease placing new orders, subject to Congressional approval, for a final procurement tally of 187 operational aircraft. The National Defense Authorization Act for Fiscal Year 2010 lacked funding for further F-22 production. The final F-22 rolled off the assembly line on 13 December 2011 during a ceremony at Dobbins Air Reserve Base.

Since 2010 the F-22 has been plagued by unresolved problems with its pilot oxygen systems which contributed to one crash and death of a pilot. The fleet was grounded for four months in 2011. The Raptor fleet has resumed flight operations, although problems with the oxygen systems have continued to be reported.

Lockheed F-117 Nighthawk

The Lockheed F-117 Nighthawk was a single-seat, twin-engine stealth ground-attack aircraft formerly operated by the United States Air Force (USAF). The F-117A's first flight was in 1981, and it achieved initial operating capability status in October 1983. The F-117A was "acknowledged" and revealed to the world in November 1988.

A product of Lockheed Skunk Works and a development of the Have Blue technology demonstrator, it became the first operational aircraft initially designed around stealth technology. The F-117A was widely publicized during the Persian Gulf War of 1991. It was commonly called the "Stealth Fighter" although it was a ground-attack aircraft, making its F-designation misleading.

The Air Force retired the F-117 on 22 April 2008, primarily because of the fielding of the F-22 Raptor and the impending introduction of the F-35 Lightning II. Sixty-four F-117s were built, 59 of which were production versions with five demonstrators/prototypes.

Glider Aircraft Variants

LET TG-10

The TG-10 is the military designation for the Blanik sailplanes used for basic flight training at the United States Air Force Academy. The Academy maintains an inventory of 21 TG-10s in three variants. The aircraft are flown by cadets and officers of the 94th Flying Training Squadron, 306th Flying Training Group, Nineteenth Air Force, Air Education and Training Command.

All of the TG-10 models are of aluminium semi-monocoque construction with fabric-covered control surfaces. All are equipped with full soaring instrument panels (altimeter, airspeed indicator, accelerometer, variometer, vertical velocity indicator, magnetic compass) and feature a full avionics suite (VHF radio, GPS, navigation computer, ELT).

TG-10B Merlin: LET L-23 Super Blanik. 12 in inventory. Basic trainer; 2-seat tandem configuration. Used in the Academy's Soar For All Program and for training cadets to become glider instructor pilots. Four of them have been configured for high altitude wave soaring.

TG-10C Kestrel/"Saber": LET L-13AC Blanik. 5 in inventory. Advanced trainer; cockpit and controls are identical to the Merlin making transitions between the two aircraft very seamless. Used for aerobatics and spin training. Slightly heavier with shorter wingspan and conventional tail configuration offers slightly faster dynamic response to control inputs.

TG-10D Peregrine/"Thunder": Blanik L-33 Solo. 4 in inventory. Advanced trainer; single seater. Cockpit and controls are similar to TG-10B. Used for advanced cross country and wave soaring training.

The Air Force Academy has retired the TG-10D sailplanes in favor of their newest high-performance gliders, the Schempp-Hirth Duo Discus and Discus 2b, designated the TG-15A (tandem two-seater) and TG-15B (single seat). In 2011, the Air Force Academy began retiring its remaining TG-10B and TG-10C gliders. Both variants are being replaced by the TG-16A.

Schweizer SGM 2-37

The Schweizer SGM 2-37 is a two-place, side-by-side, fixed gear, low wing motor glider.

A total of twelve were produced between 1982 and 1988, including nine for the United States Air Force Academy, which designated it the TG-7A. The TG-7A was retired from USAFA service in April 2003.

The basic airframe was later developed into the SA 2-37A and B covert surveillance aircraft.

Search and Rescue Aircraft Variants

Bell UH-1 Iroquois variants

The Bell UH-1 Iroquois is a military helicopter powered by a single, turboshaft engine, with a two-bladed main rotor and tail rotor.

The helicopter was developed by Bell Helicopter to meet the United States Army's requirement for a medical evacuation and utility helicopter in 1952, and first flew on 20 October 1956.

Ordered into production in March 1960, the UH-1 was the first turbine-powered helicopter to enter production for the United States military, and more than 16,000 have been produced worldwide.

The first combat operation of the UH-1 was in the service of the U.S. Army during the Vietnam War.

The original designation of HU-1 led to the helicopter's nickname of Huey. In September 1962, the designation was changed to UH-1, but "Huey" remained in common use. Approximately 7,000 UH-1 aircraft saw service in Vietnam.

In 1952, the Army identified a requirement for a new helicopter to serve as medical evacuation (MEDEVAC), instrument trainer and general utility aircraft. The Army determined that current helicopters were too large, underpowered, or were too complex to maintain easily.

In November 1953, revised military requirements were submitted to the Department of the Army. Twenty companies submitted designs in their bid for the contract, including Bell Helicopter with the Model 204 and Kaman Aircraft with a turbine-powered version of the H-43. On 23 February 1955, the Army announced its decision, selecting Bell to build three copies of the Model 204 for evaluation, designated as the XH-40.

Test Aircraft Variants

A-6 Intruder

A U.S. Navy Grumman A-6 Intruder (BuNo 155670) aircraft from attack squadron VA-52 Knightriders in 1981.

VA-52 was assigned to Carrier Air Wing 15 (CVW-15) aboard the aircraft carrier USS Kitty Hawk (CV-63) from 1 April to 23 November 1981 for a deployment to the Western Pacific.

The date given (15 January 1984) has to be incorrect, as BuNo 155670 was assigned to VA-52 for the 1981 cruise. In 1984 CVW-15 was also assigned to the USS Carl Vinson (CVN-70). 155670 was an A-6A modified to an A-6E.

MD Helicopters MD Explorer

The MD Helicopters MD Explorer is a light twin utility helicopter. Designed in the early 1990s by McDonnell Douglas Helicopter Systems, it is currently produced by MD Helicopters, Inc. There have been two models, the original MD 900, and its successor, the MD 902.

In January 1989, McDonnell Douglas Helicopters officially launched the development of the Explorer, initially referred to as MDX.

The Explorer was the first McDonnell Douglas helicopter to incorporate the NOTAR system from its initial design.

McDonnell Douglas partnered with Hawker de Havilland of Australia to manufacture the airframes. 10 prototypes were built with seven being used for ground tests.

McDonnell Douglas Helicopters became a launch customer for Pratt & Whitney Canada's PW200 series of engines, with an exclusive agreement to power the first 128 Explorers with two PW206As. Meanwhile, plans to offer the Turbomeca Arrius as an option were dropped.

The first flight of the Explorer took place on 18 December 1992, with ship #2 (N900MD). FAA certification for the Explorer was granted on 2 December 1994, with JAA certification following shortly after.

The MD Explorer features the NOTAR anti-torque system, with benefits including increased safety, far lower noise levels and performance and controllability enhancements. Instead of an anti-torque tail rotor, a fan exhaust is directed out slots in the tail boom, using the Coandă effect for yaw control. Boeing retains the design rights to the NOTAR technology despite selling the former McDonnell Douglas civil helicopter line to MD Helicopters in early 1999. The Explorer also features an advanced bearingless five blade main rotor with composite blades, plus carbonfibre construction tail and fuselage.

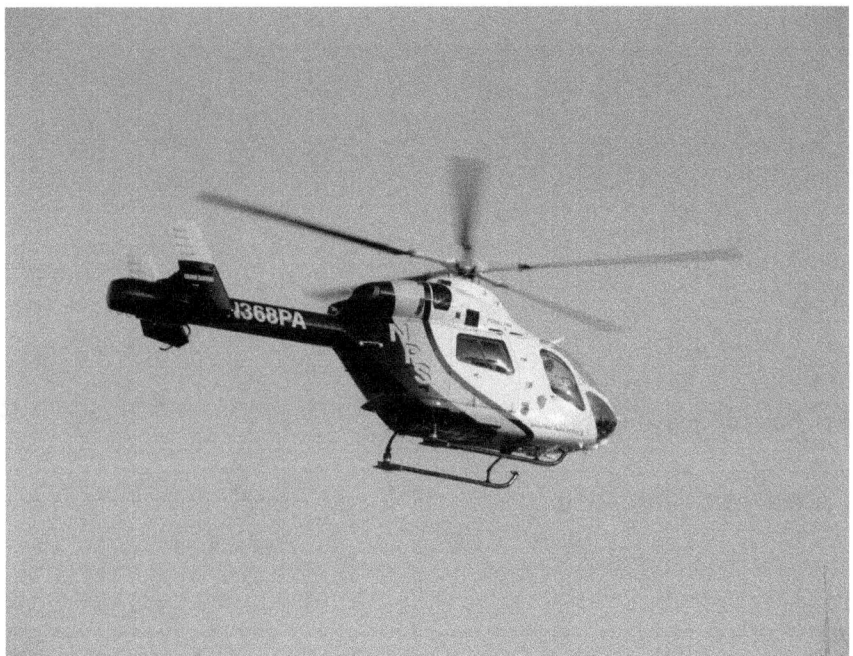

Patrol Aircraft Variants

Lockheed P-3 Orion

The Lockheed P-3 Orion is a four-engine turboprop anti-submarine and maritime surveillance aircraft developed for the United States Navy and introduced in the 1960s. Lockheed based it on the L-188 Electra commercial airliner. The aircraft is easily recognizable by its distinctive tail stinger or "MAD Boom", used for the magnetic detection of submarines.

Over the years, the aircraft has seen numerous design advancements, most notably to its electronics packages. The P-3 Orion is still in use by numerous navies and air forces around the world, primarily for maritime patrol, reconnaissance, anti-surface warfare and anti-submarine warfare. A total of 734 P-3s have been built, and by 2012, it will join the handful of military aircraft such as the Boeing B-52 Stratofortress which have served 50 years of continuous use with its original primary customer, in this case, the United States Navy. The U.S. Navy's remaining P-3C aircraft will eventually be replaced by the Boeing P-8A Poseidon.

Reconnaissance Aircraft Variants

Lockheed SR-71 Blackbird

The Lockheed SR-71 "Blackbird" was an advanced, long-range, Mach 3+ strategic reconnaissance aircraft. It was developed as a black project from the Lockheed A-12 reconnaissance aircraft in the 1960s by the Lockheed Skunk Works. Clarence "Kelly" Johnson was responsible for many of the design's innovative concepts. During reconnaissance missions the SR-71 operated at high speeds and altitudes to allow it to outrace threats. If a surface-to-air missile launch was detected, the standard evasive action was simply to accelerate and outrun the missile.

The SR-71 served with the U.S. Air Force from 1964 to 1998. Of the 32 aircraft built, 12 were destroyed in accidents, and none were lost to enemy action. The SR-71 has been given several nicknames, including Blackbird and Habu, the latter in reference to an Okinawan species of pit viper. Since 1976, it has held the world record for the fastest air-breathing manned aircraft, a record previously held by the YF-12.

The SR-71 was designed to minimize its radar cross-section, an early attempt at stealth design.

The high temperatures generated by Mach 3 flight required its airframe to be made mostly of titanium. To control costs, Lockheed used a more easily worked alloy of titanium which softened at a lower temperature.

Finished aircraft were painted a dark blue, almost black, to increase the emission of internal heat and to act as camouflage against the night sky. The dark color led to the aircraft's call sign "Blackbird".

Anti-Submarine Aircraft Variants

Grumman S-2 Tracker

The Grumman S-2 Tracker (previously S2F prior to 1962) was the first purpose-built, single airframe anti-submarine warfare (ASW) aircraft to enter service with the US Navy. The Tracker was of conventional design with twin engines, a high wing and tricycle undercarriage. The type was exported to a number of navies around the world. Introduced in 1952 the Tracker saw service in the USN until the mid-1970s with a few aircraft remaining in service with other air arms into the 21st century. The last operating fleet is maintained by Argentina and Brazil.

Intended as a replacement for its predecessor, Grumman's AF-2 Guardian which was the first purpose-built aircraft system for ASW, using two airframes, one with the detection gear, and the other with the weapon systems, the Tracker combined both functions in one aircraft. Grumman's design (model G-89) was for a large high-wing monoplane with twin Wright Cyclone R-1820 nine cylinder radial engines, a yoke type arrestor hook and a crew of four. Both the two prototypes XS2F-1 and 15 production aircraft, S2F-1 were ordered at the same time, on 30 June 1950. The first flight was conducted on 4 December 1952, and production aircraft entered service with VS-26, in February 1954.

Follow-on versions included the WF Tracer and TF Trader, which became the Grumman E-1 Tracer and Grumman C-1 Trader in the tri-service designation standardization of 1962. The S-2 carried the nickname "Stoof" (S-two-F) throughout its military career; and the E-1 Tracer variant with the large overhead radome was colloquially called the "stoof with a roof."

Grumman produced 1,185 Trackers. Another 99 aircraft carrying the CS2F designation were manufactured in Canada under license by de Havilland Canada. U.S.-built versions of the Tracker were sold to various nations, including Australia, Japan, Turkey and Taiwan.

Utility Aircraft Variants

Lockheed U-2

The Lockheed U-2, nicknamed "Dragon Lady", is a single-engine, very high-altitude reconnaissance aircraft operated by the United States Air Force (USAF) and previously flown by the Central Intelligence Agency (CIA). It provides day and night, very high-altitude (70,000 feet / 21,000 m), all-weather intelligence gathering. The aircraft is also used for electronic sensor research and development, satellite calibration, and satellite data validation.

The unique design that gives the U-2 its remarkable performance also makes it a difficult aircraft to fly. It was designed and manufactured for minimum airframe weight, which results in an aircraft with little margin for error. Most aircraft were single-seat versions, with only five two-seat trainer versions known to exist. Early U-2 variants were powered by Pratt & Whitney J57 turbojet engines. The U-2C and TR-1A variants used the more powerful Pratt & Whitney

J75 turbojet. The U-2S and TU-2S variants incorporated the even more powerful General Electric F118 turbofan engine.

High-aspect-ratio wings give the U-2 some glider-like characteristics, with a lift-to-drag ratio estimated in the high 20s. To maintain their operational ceiling of 70,000 feet (21,000 m), the U-2A and U-2C models (no longer in service) must fly very near their maximum speed. The aircraft's stall speed at that altitude is only 10 knots (12 mph; 19 km/h) below its maximum speed. This narrow window was referred to by the pilots as the "coffin corner". For 90% of the time on a typical mission the U-2 was flying within only five knots above stall, which might cause a decrease in altitude likely to lead to detection, and additionally might overstress the lightly built airframe.

Experimental Aircraft Variants

Northrop X-4 Bantam

The Northrop X-4 Bantam was a prototype small twin-jet aircraft manufactured by Northrop Corporation in 1948. It had no horizontal tail surfaces, depending instead on combined elevator and aileron control surfaces (called elevons) for control in pitch and roll attitudes, almost exactly in the manner of the similar-format, rocket-powered Messerschmitt Me 163 of Nazi Germany's Luftwaffe. Some aerodynamicists had proposed that eliminating the horizontal tail would also do away with stability problems at fast speeds (called shock stall) resulting from the interaction of supersonic shock waves from the wings and the horizontal stabilizers. The idea had merit, but the flight control systems of that time prevented the X-4 from any success.

Two X-4s were built by the Northrop Corporation, but the first was found to be mechanically unsound and after 10 flights it was grounded and used to provide parts for the second. While being tested from 1950 to 1953 at the NACA High-Speed Flight Research Station (now Edwards Air Force Base), the X-4's semi-tailless configuration exhibited inherent longitudinal stability problems (porpoising) as it approached the speed of sound. It was concluded that (with the control technology available at the time) tailless craft were not suited for transonic flight.

It was believed in the 1940s that a design without horizontal stabilizers would avoid the interaction of shock waves between the wing and stabilizers. These were believed to be the source of the stability problems at transonic speeds up to Mach 0.9. Two aircraft had already been built using a semi-tailless design—the rocket-powered Me 163 Komet flown by Germany in World War II, and the British de Havilland DH.108 Swallow built after the war. The United States Army Air Forces signed a contract with the Northrop Aircraft Company on 11 June 1946, to build two X-4s. Northrop was selected because of its experience with flying wing designs, such as the N-9M, XB-35 and YB-49 aircraft.

The resulting aircraft was very compact, only large enough to hold two Westinghouse J30 jet engines, a pilot, instrumentation, and a 45-minute fuel supply. Nearly all maintenance work on the aircraft could be done without using a ladder or footstool. A person standing on the ground could easily look into the cockpit. The aircraft also had split flaps, which doubled as speed brakes.

The first X-4 (serial number 46-676) was delivered to Muroc Air Force Base, California, in November 1948. It underwent taxi tests and made its first flight on December 15, 1948, with Northrop test pilot Charles Tucker at the controls. Winter rains flooded Rogers Dry Lake soon after, preventing additional X-4 flights until April 1949. The first X-4 proved mechanically unreliable, and made only 10 flights. Walt Williams, the head of the NACA Muroc Flight Test Unit (now Dryden Flight Research Center) called the aircraft a "lemon". The second X-4 (serial number 46-677) was delivered during the halt of flights, and soon proved far more reliable. It made a total of 20 contractor flights. Despite this, the contractor flight program dragged on until February 1950, before both aircraft were turned over to the Air Force and the NACA. The first X-4 never flew again, used as spare parts for the second aircraft.

The NACA instrumented the second X-4 to conduct a short series of flights with Air Force pilots. These included Chuck Yeager, Frank Kendall Everest, Jr., Al Boyd, Richard Johnson, Fred Ascani, Arthur Murray and Jack Ridley. The flights were made in August and September 1950. The first flight by a NACA pilot was made by John H. Griffith on September 28, 1950.

The initial NACA X-4 flights, which continued from late 1950 through May of 1951, focused on the aircraft's sensitivity in pitch. NACA pilots Griffith and Scott Crossfield noted that as the X-4's speed approached Mach 0.88, it began a pitch oscillation of increasing severity, which was likened to driving on a washboard road. Increasing speeds also caused a tucking phenomena, in which the nose pitched down, a phenomenon also experienced by the Me 163A Anton prototypes in 1941. More seriously, the aircraft also showed a tendency to "hunt" about all three axes. This combined yaw, pitch and roll, which grew more severe as the speed increased, was a precursor to the inertial coupling which would become a major challenge in the years to come.

To correct the poor stability, project engineers decided to increase the flap/speed brake trailing edge thickness. Balsa wood strips were added between the flap/speed brake halves, causing them to remain open at a 5° angle. The first test of the blunt trailing edge was flown on 20 August 1951, by NACA pilot Walter Jones. A second test was made by Crossfield in October. The results were positive, with Jones commenting that the X-4's flight qualities had been greatly improved, and the aircraft did not have pitch control problems up to a speed of Mach 0.92.

The balsa strips were removed, and the X-4 then undertook a long series of flights to test landing characteristics. By opening the speed brakes, the lift-to-drag ratio of the aircraft could be reduced to less than 3:1. This was for data on future rocket-powered aircraft. The tests continued through October 1951, until wing tank fuel leaks forced the aircraft to be grounded until March 1952, when the landing tests resumed. NACA pilots Joe Walker, Stanley Butchard, and George Cooper were also checked out in the aircraft.

The thickened flap/speed brake tests had been encouraging, so balsa wood strips were reinstalled on both the flap/speed brake and the elevons. The first flight was made by Jones on 19 May 1952, but one of the engines was damaged during the flight, and it was August before a replacement J30 could be found. When the flights resumed, they showed that the modifications had improved stability in both pitch and yaw, and delayed the nosedown trim changes from Mach 0.74 to Mach 0.91. Above Mach 0.91, however, the X-4 still oscillated.

In May 1953, the balsa wood strips were again removed, and the X-4's dynamic stability was studied in the original flap/speed brake and elevon configuration. These flights were made by Crossfield and John B. McKay. This was the final project for the X-4, which made its 81st and final NACA flight on September 29, 1953. Both aircraft survived the test program. The first X-4 was transferred to the United States Air Force Academy, Colorado Springs, Colorado, before being returned to Edwards Air Force Base. The second X-4 went to the National Museum of the United States Air Force at Wright-Patterson Air Force Base near Dayton, Ohio, where it remains on display.

The X-4's primary importance involved proving a negative, in that a swept-wing semi-tailless design was not suitable for speeds near Mach 1, although the F7U Cutlass proved to be something of a counterexample—the developed version was the first aircraft to demonstrate stores separation above Mach 1. Aircraft designers were thus able to avoid this dead end. It was not until the development of computer fly-by-wire systems that such designs could be practical. Semi-tailless designs appeared on the X-36, Have Blue, F-117, and Bird of Prey, although these aircraft all differed significantly in shape from the X-4. The trend during its test program was already toward delta and modified delta aircraft such as the Douglas F4D, the Convair F-102A derived from the XF-92A, and the Avro Vulcan.

Convair X-6

The Convair X-6 was a proposed experimental aircraft project to develop and evaluate a nuclear-powered jet aircraft. The project was to use a Convair B-36 bomber as a testbed aircraft, and though one NB-36H was modified during the early stages of the project, the program was cancelled before the actual X-6 and its nuclear reactor engines were completed. The X-6 was part of a larger series of programs, costing US$7 billion in all, that ran from 1946 through 1961. Because such an aircraft's range would not have been limited by liquid jet fuel, it was theorized that nuclear-powered strategic bombers would be able to stay airborne for weeks at a time.

In May, 1946, the Nuclear Energy for the Propulsion of Aircraft (NEPA) project was started by the Air Force. Studies under this program were done until May, 1951 when NEPA was replaced by the Aircraft Nuclear Propulsion (ANP) program. The ANP program contained plans for two B-36s to be modified by Convair under the MX-1589 project. One of the B-36s was to be used to study shielding requirements for an airborne reactor while the other was to be the X-6.

The first modified B-36 was called the Nuclear Test Aircraft (NTA), a B-36H-20-CF (Serial Number 51-5712) that had been damaged in a tornado at Carswell AFB on September 1, 1952. This plane was redesignated the XB-36H, then the NB-36H and was modified to carry a 3 megawatt, air-cooled nuclear reactor in its bomb bay. The reactor, named the Aircraft Shield Test Reactor (ASTR), was operational but did not power the plane. Water, acting as both moderator and coolant, was pumped through the reactor core and then to water-to-air heat exchangers to dissipate the heat to the atmosphere. Its sole purpose was to investigate the effect of radiation on aircraft systems.

To shield the flight crew, the nose section of the aircraft was modified to include a 12-ton lead and rubber shield. The standard windshield was replaced with one made of 6-inch-thick (15 cm) acrylic glass. The amount of lead and water shielding was variable. Measurements of the resulting radiation levels were then compared with calculated levels to enhance the ability to design optimal shielding with minimum weight for nuclear-powered bombers.

The NTA completed 47 test flights and 215 hours of flight time (during 89 of which the reactor was operated) between September 17, 1955, and March 1957 over New Mexico and Texas. This was the only known airborne reactor experiment by the U.S. with an operational nuclear reactor on board. The NB-36H was scrapped at Fort Worth in 1958 when the Nuclear Aircraft Program was abandoned. After the ASTR was removed from the NB-36H, it was moved to the National Aircraft Research Facility.

Thrust vectoring

Thrust vectoring, also thrust vector control or TVC, is the ability of an aircraft, rocket, or other vehicle to manipulate the direction of the thrust from its engine(s) or motor in order to control the attitude or angular velocity of the vehicle.

119

In rocketry and ballistic missiles that fly outside the atmosphere, aerodynamic control surfaces are ineffective, so thrust vectoring is the primary means of attitude control.

For aircraft, the method was originally envisaged to provide upward vertical thrust as a means to give aircraft vertical (VTOL) or short (STOL) takeoff and landing ability. Subsequently, it was realized that using vectored thrust in combat situations enabled aircraft to perform various maneuvers not available to conventional-engined planes. To perform turns, aircraft that use no thrust vectoring must rely on aerodynamic control surfaces only, such as ailerons or elevator; craft with vectoring must still use control surfaces, but to a lesser extent.

Thrust vector control is effective only while the propulsion system is creating thrust. At other stages of flight, separate mechanisms are required for attitude and flight path control.

Nominally, the line of action of the thrust vector of a rocket nozzle passes through the vehicle's center of mass, generating zero net moment about the mass center. It is possible to generate pitch and yaw moments by deflecting the main rocket thrust vector so that it does not pass through the mass center. Because the line of action is generally oriented nearly parallel to the roll axis, roll control usually requires the use of two or more separately hinged nozzles or a separate system altogether, such as fins.

Thrust vectoring for many liquid rockets is achieved by gimballing the rocket engine. This often involves moving the entire combustion chamber and outer engine bell, or even the entire engine assembly including the related fuel and oxidizer pumps. Such a system was used on the Saturn V and the Space Shuttle.

Another method of thrust vectoring used on early solid propellant ballistic missiles was liquid injection, in which the rocket nozzle is fixed, but a fluid is introduced into the exhaust flow from injectors mounted around the aft end of the missile. If the liquid is injected on only one side of the missile, it modifies that side of the exhaust plume, resulting in different thrust on that side and an asymmetric net force on the missile. This was the control system used on the Minuteman II and the early SLBMs of the United States Navy.

A later method developed for solid propellant ballistic missiles achieves thrust vectoring by deflecting the rocket nozzle using electric servomechanisms or hydraulic cylinders. The nozzle is attached to the missile via a ball joint with a hole in the center, or a flexible seal made of a thermally resistant material, the latter generally requiring more torque and a higher power actuation system. The Trident C4 and D5 systems are controlled via hydraulically actuated nozzle.

Some smaller sized atmospheric tactical missiles, such as the AIM-9X Sidewinder, eschew flight control surfaces and instead use mechanical vanes to deflect motor exhaust to one side.

A famous example of thrust vectoring is the Rolls-Royce Pegasus engine used in the Hawker Siddeley Harrier, as well as in the AV-8B Harrier II variant.

Widespread use of thrust vectoring for enhanced maneuverability in Western production-model fighter aircraft would have to wait until the 21st century, and the deployment of the Lockheed Martin F-22 Raptor fifth-generation jet fighter, with its afterburning, thrust-vectoring Pratt & Whitney F119 turbofan.

The Harrier—the world's first operational fighter jet with thrust vectoring, enabling VTOL capabilities

Lockheed Martin F-35 Lightning II is currently in the pre-production test and development stage.

The Lockheed Martin F-35 Lightning II is a family of single-seat, single-engine, fifth generation multirole fighters under development to perform ground attack, reconnaissance, and air defense missions with stealth capability. The F-35 has three main models; the F-35A is a conventional takeoff and landing variant, the F-35B is a short take off and vertical-landing variant, and the F-35C is a carrier-based variant.

The F-35 is descended from the X-35, the product of the Joint Strike Fighter (JSF) program. JSF development is being principally funded by the United States, with the United Kingdom and other partner governments providing additional funding. The partner nations are either NATO members or close U.S. allies. It is being designed and built by an aerospace industry team led by Lockheed Martin. The F-35 carried out its first flight on 15 December 2006.

Although this aircraft uses a conventional afterburning turbofan (F135 or F136) to facilitate supersonic operation, the F-35B variant, developed for joint usage by the US Marine Corps, UK Royal Air Force and Royal Navy, also incorporates a vertically mounted, low-pressure shaft-driven remote fan, which is driven through a clutch during landing from the engine. Both the exhaust from this fan and the main engine's fan are deflected by thrust vectoring nozzles, to provide the appropriate combination of lift and propulsive thrust.

Bell X-22

The Bell X-22 was a United States V/STOL X-plane with four tilting ducted fans. Take-off was to selectively occur either with the propellers tilted vertically upwards, or on a short runway with the nacelles tilted forward at approximately 45°. Additionally, the X-22 was to provide more insight into the tactical application of vertical take-off troop transporters such as the preceding Hiller X-18 and the X-22 successor, the Bell XV-15. Another program requirement was a true airspeed in level flight of at least 525 km/h.

The maiden flight of the prototype occurred on 17 March 1966. In contrast to other tilt-rotor craft (such as the Bell XV-3), transitions between hovering and horizontal flight succeeded nearly immediately. However, interest increased more towards VTOL and V/STOL properties, not the specific design of the prototype.

The Bell XV-3 (Bell 200) was a tiltrotor aircraft developed by Bell Helicopter for a joint research program between the United States Air Force and the United States Army, to explore convertiplane technologies. The XV-3 featured an engine mounted in the fuselage with drive shafts transferring power to, two-bladed rotor assemblies mounted on the wingtips. The wingtip rotor assemblies were mounted to tilt 90 degrees from vertical to horizontal, designed to allow the XV-3 to take off and land like a helicopter but fly at faster airspeeds, similar to a conventional fixed-wing aircraft.

Martin Marietta X-24A

The X-24 was one of a group of lifting bodies flown by the NASA Flight Research Center (now Dryden Flight Research Center) in a joint program with the U.S. Air Force at Edwards Air Force Base in California from 1963 to 1975. The lifting bodies were used to demonstrate the ability of pilots to maneuver and safely land wingless vehicles designed to fly back to Earth from space and be landed like an airplane at a predetermined site.

Lifting bodies' aerodynamic lift, essential to flight in the atmosphere, was obtained from their shape. The addition of fins and control surfaces allowed the pilots to stabilize and control the vehicles and regulate their flight paths.

The X-24A was a fat, short teardrop shape with vertical fins for control. It made its first, unpowered, glide flight on April 17, 1969 with Air Force Maj. Jerauld

R. Gentry at the controls. Gentry also piloted its first powered flight on March 19, 1970. The craft was taken to around 45,000 feet (13.7 km) by a modified B-52 and then drop launched, then either glided down or used its rocket engine to ascend to higher altitudes before gliding down. The X-24A was flown 28 times at speeds up to 1,036 mph (1,667 km/h) and altitudes up to 71,400 feet (21.8 km).The X-24A was a fat, short teardrop shape with vertical fins for control. It made its first, unpowered, glide flight on April 17, 1969 with Air Force Maj. Jerauld

R. Gentry at the controls. Gentry also piloted its first powered flight on March 19, 1970. The craft was taken to around 45,000 feet (13.7 km) by a modified B-52 and then drop launched, then either glided down or used its rocket engine to ascend to higher altitudes before gliding down. The X-24A was flown 28 times at speeds up to 1,036 mph (1,667 km/h) and altitudes up to 71,400 feet (21.8 km).

The Martin Marietta X-24A was an experimental US aircraft developed from a joint USAF-NASA program named PILOT (1963–1975). It was designed and built to test lifting body concepts, experimenting with the concept of unpowered reentry and landing, later used by the Space Shuttle.

Bensen B-8

The Bensen B-8 is a small, single-seat autogyro developed in the United States in the 1950s. Although the original manufacturer stopped production in 1987, plans for homebuilders are still available as of 2007. Its design was a refinement of the Bensen B-7, and like that aircraft, the B-8 was initially built as an unpowered rotor-kite. It first flew in this form in 1955, and on 6 December a powered version, designated B-8M (M for motorised) first flew. The design proved to be extremely popular and long-lasting, with thousands of sets of plans sold over the next thirty years.

Rockwell X-30

The Rockwell X-30 was an advanced technology demonstrator project for the National Aero-Space Plane (NASP), part of a United States project to create a single-stage-to-orbit (SSTO) spacecraft and passenger spaceliner. It was cancelled in the early 1990s, before a prototype was completed, although a lot of development work in advanced materials and aerospace design was completed. While a goal of a future NASP was a passenger liner capable of two hour "trips" from Washington to Tokyo, the X-30 was planned for a crew of two and oriented towards testing.

The NASP concept is thought to have come from the "Copper Canyon" project, in Defense Advanced Research Projects Agency (DARPA), running from 1982 to 1985. In his 1986 State of the Union address, President Ronald Reagan called for "a new Orient Express that could, by the end of the next decade, take off from Dulles Airport, accelerate up to 25 times the speed of sound, attaining low earth orbit or flying to Tokyo within two hours."

Research suggested a maximum speed of Mach 8 for scramjet based aircraft, as the vehicle would generate heat due to atmospheric friction, which would thus cost considerable energy. The project showed that much of this energy could be recovered by passing hydrogen over the skin and carrying the heat into the combustion chamber: Mach 20 then seemed possible. The result was a program funded by NASA, and the United States Department of Defense (funding was approximately equally divided between NASA, DARPA, the US Air Force, the Strategic Defense Initiative Office (SDIO) and the US Navy).

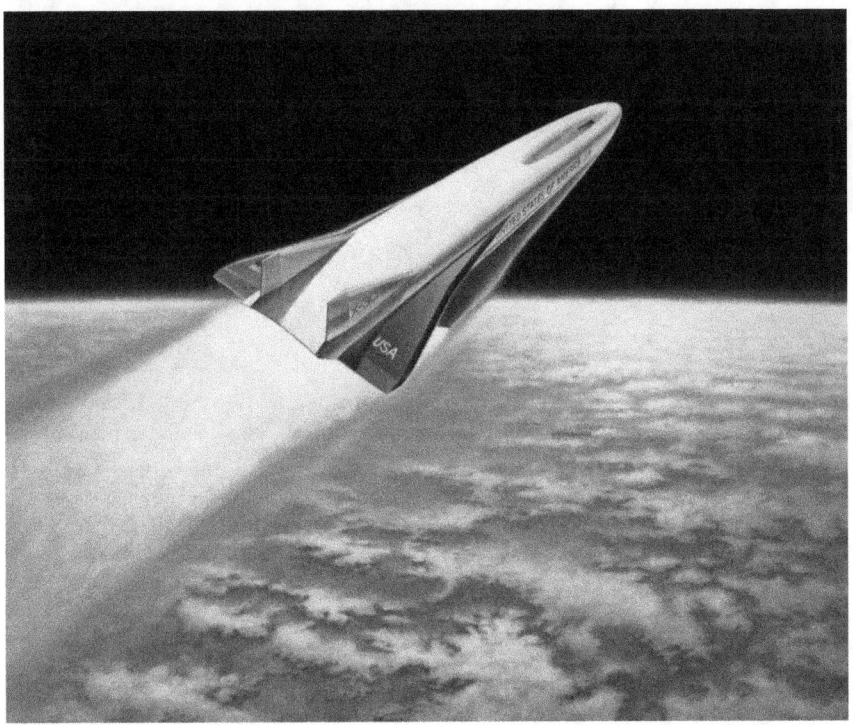

Rockwell-MBB X-31

The collaborative Rockwell-Messerschmitt-Bölkow-Blohm X-31 Enhanced Fighter Maneuverability program was designed to test fighter thrust vectoring technology. Thrust vectoring allows the X-31 to fly in a direction other than where the nose is pointing, resulting in significantly more maneuverability than most conventional fighters. An advanced flight control system provides controlled flight at high angles of attack where conventional aircraft would stall.

Two X-31s were built, with the first flying on October 11, 1990. Over 500 test flights were carried out between 1990 and 1995. The X-31 featured fixed strakes along the aft fuselage, as well as a pair of movable computer-controlled canards to increase stability and maneuverability. There are no horizontal tail surfaces, only the vertical fin with rudder. Pitch and yaw are controlled by the three paddles directing the exhaust (thrust vectoring). Eventually, simulation tests on one of the X-31s showed that flight would have been stable had the plane been designed without the vertical fin, because the thrust-vectoring nozzle provided sufficient yaw and pitch control.

Joint Strike Fighter program

Joint Strike Fighter (JSF) is a development and acquisition program intended to replace a wide range of existing fighter, strike, and ground attack aircraft for the United States, the United Kingdom, Canada, Australia and their allies. After a competition between the Boeing X-32 and the Lockheed Martin X-35, a final design was chosen based on the X-35. This is the F-35 Lightning II, which will replace various tactical aircraft, including the US F-16, A-10, F/A-18, AV-8B and British Harrier GR7 & GR9s, and the Canadian CF-18. The projected average annual cost of this program is $12.5 billion with an estimated program life-cycle cost of $1.1 trillion.

The JSF program was the result of the merger of the Common Affordable Lightweight Fighter (CALF) and Joint Advanced Strike Technology (JAST) projects. The merged project continued under the JAST name until the engineering, manufacturing and development (EMD) phase, during which the project became the Joint Strike Fighter.

The CALF was an ARPA program to develop a STOVL strike fighter (SSF) for the United States Marine Corps and replacement for the F-16 Fighting Falcon. The United States Air Force passed over the F-16 Agile Falcon in the late 1980s, essentially an enlarged F-16, and continued to mull other designs. In 1992 the Marine Corps and Air Force agreed to jointly develop the Common Affordable Lightweight Fighter, also known as Advanced Short Takeoff and Vertical Landing (ASTOVL), after Paul Bevilaqua persuaded the Air Force that his team's concept had potential as an F-22 complement, stripped of the lift system. Thus in a sense the F35B begat the F35A, not the other way around.

The Boeing X-32 was a multi-purpose jet fighter in the Joint Strike Fighter contest.

It lost to the Lockheed Martin X-35 demonstrator which was further developed into the Lockheed Martin F-35 Lightning II.

Lockheed Martin X-33

The Lockheed Martin X-33 was an unmanned, sub-scale technology demonstrator suborbital spaceplane developed in the 1990s under the U.S. government-funded Space Launch Initiative program. The X-33 was a technology demonstrator for the VentureStar orbital spaceplane, which was planned to be a next-generation, commercially operated reusable launch vehicle. The X-33 would flight-test a range of technologies that NASA believed it needed for single-stage-to-orbit reusable launch vehicles (SSTO RLVs), such as metallic thermal protection systems, composite cryogenic fuel tanks for liquid hydrogen, the aerospike engine, autonomous (unmanned) flight control, rapid flight turn-around times through streamlined operations, and its lifting body aerodynamics.

Failures of the enormous, multi-lobe composite material fuel cells during pressure testing ultimately led to the cancellation of the program as a federal program in 2001, but Lockheed Martin has conducted related testing, and has had successes as recently as 2009.

Orbital Sciences X-34

The Orbital Sciences X-34 was intended as a low-cost testbed to demonstrate "key technologies" integratable to the Reusable Launch Vehicle program.

It was intended to be an autonomous pilotless craft powered by a 'Fastrac' liquid rocket engine capable of reaching Mach 8, and performing 25 test flights per year.

The unpowered prototype had only been used for towing and captive flight tests when the project was canceled in 2001 for cost concerns. Orbital and Rockwell withdrew less than a year after the contract was signed, because they decided the project could not be done for the promised amount. (A major disagreement between Rockwell and NASA over engine choice likely contributed to the decision.)

The X-34 was reborn as a program for a suborbital reusable-rocket technology demonstrator. But, in early 2001, the first flight vehicle was near completion, the program closed after NASA demanded sizable design changes without further funding. The contractor, Orbital Sciences, refused. To this point, the project had encompassed spending of just under $112 million: $85.7M from the original contract with designer Orbital Sciences, $16M from NASA and various government agencies for testing, and an additional $10M for Orbital Sciences to adapt its L-1011 carrier to accommodate the X-34. The program was officially cancelled by NASA on March 1, 2001.

The two demonstrators remained in storage at Edwards Air Force Base until November 16, 2010, when both X-34s were moved with their vertical tails removed from Dryden to a hangar owned by the National Test Pilot school in Mojave, California. They are to be inspected, and NASA is investigating the possibility of restoring them to flight status.

Boeing X-37

The Boeing X-37 (also known as the X-37 Orbital Test Vehicle) is an American reusable unmanned spacecraft. It is boosted into space by a rocket, then re-enters Earth's atmosphere and lands as a spaceplane. The X-37 is

131

operated by the United States Air Force for orbital spaceflight missions intended to demonstrate reusable space technologies. It is a 120%-scaled derivative of the earlier Boeing X-40.

The X-37 began as a NASA project in 1999, before being transferred to the U.S. Department of Defense in 2004. It conducted its first flight as a drop test on 7 April 2006, at Edwards Air Force Base, California. The spaceplane's first orbital mission, USA-212, was launched on 22 April 2010 using an Atlas V rocket. Its successful return to Earth on 3 December 2010 was the first test of the vehicle's heat shield and hypersonic aerodynamic handling. A second X-37 was launched on 5 March 2011, with the mission designation USA-226; it returned to Earth on 16 June 2012.

The vehicle that was used as an atmospheric drop test glider had no propulsion system. Instead of an operational vehicle's payload bay doors, it had an enclosed and reinforced upper fuselage structure to allow it to be mated with a mothership. In September 2004, DARPA announced that for its initial atmospheric drop tests the X-37 would be launched from the Scaled Composites White Knight, a high-altitude research aircraft.

On 21 June 2005, the X-37A completed a captive-carry flight underneath the White Knight from Mojave Spaceport in Mojave, California. Through the second half of 2005, the X-37A underwent structural upgrades, including the reinforcement of its nose wheel supports. Further captive-carry flight tests and the first drop test were initially expected to occur in mid-February 2006. The X-37's public debut was scheduled for its first free flight on 10 March 2006, but was canceled due to an Arctic storm. The next flight attempt, on 15 March 2006, was canceled due to high winds.

On 24 March 2006, the X-37 flew again, but a datalink failure prevented a free flight, and the vehicle returned to the ground still attached to its White Knight carrier aircraft.

On 7 April 2006, the X-37 made its first free glide flight. During landing, the vehicle overran the runway and sustained minor damage. Following the vehicle's extended downtime for repairs, the program moved from Mojave to Air Force Plant 42 (KPMD) in Palmdale, California for the remainder of the flight test program. White Knight continued to be based at Mojave, but was ferried over to Plant 42 when flights were scheduled.

Five additional flights were performed, two of which resulted in X-37 releases with successful landings. These two free flights occurred on 18 August 2006 and 26 September 2006.

OTV-1, the first X-37B, launched on its first mission – USA-212 – on an Atlas V rocket at Cape Canaveral Air Force Station, Florida, on 22 April 2010 at 23:58 GMT.

The spacecraft was placed into low Earth orbit for testing. While the U.S. Air Force revealed few orbital details of the mission, amateur astronomers claimed to have identified the experimental spacecraft in orbit and shared their findings. A worldwide network of amateur astronomers reported that on 22 May the spacecraft was in an inclination of 39.99 degrees, circling the Earth once every 90 minutes on an orbit 249 by 262 miles (401 by 422 km).

The X-37B reputedly passed over the same given spot on Earth every four days, and operated at an altitude of 255 miles (410 km), which is typical for military surveillance satellites. However, such an orbit is also common among civilian LEO satellites, and the spaceplane's altitude was the same as that of the ISS and most other manned spacecraft.

The U.S. Air Force announced on 30 November 2010 that the X-37B would return for a landing during the 3–6 December timeframe.

As scheduled, OTV-1 de-orbited, reentered Earth's atmosphere, and landed successfully at Vandenberg AFB on 3 December 2010, at 1:16 PST (09:16 UTC), conducting America's first autonomous orbital landing onto a runway; the first spacecraft to perform such a feat was the Soviet Buran shuttle in 1988. In all, the X-37B spent 224 days in space. OTV-1 suffered a tire blowout during landing and sustained minor damage to its underside.

A second X-37B mission, designated USA-226, was launched aboard an Atlas V rocket Cape Canaveral Air Force Station, Florida on 5 March 2011.

The mission was classified and described by the U.S. military as an effort to test new space technologies. On 29 November 2011, the U.S. Air Force announced that it would extend the mission of USA-226 beyond the 270-day baseline design duration. In April 2012, General William L. Shelton of the Air Force Space Command declared the ongoing mission a "spectacular success".

On 30 May 2012, the Air Force stated that OTV-2 would complete its mission and land at Vandenberg AFB in June 2012. The spacecraft landed autonomously on 16 June 2012, having spent 469 days in space.

NASA X-38

The X-38 Crew Return Vehicle (CRV) was a prototype for a wingless lifting body reentry vehicle that was to be used as a Crew Return Vehicle for the International Space Station (ISS). The X-38 was developed to the point of a drop test vehicle before its development was cancelled in 2002 due to budget cuts.

X-38 was the program under leadership of NASA Johnson Space Center to build a series of incremental flight demonstrators for the proposed Crew Return Vehicle. In an unusual move for an X-plane, the program involved the European Space Agency and the German Space Agency DLR. It was originally called X-35. The program manager was John Muratore, while the Flight Test Engineer was future NASA astronaut Michael E. Fossum.

The X-38 design used a wingless lifting body concept originally developed by the U.S. Air Force in the mid-1960s during the X-24 program, and it was Muratore's brainchild.

Following the jettison of a deorbit engine, the X-38 would have glided from orbit and used a steerable parafoil for its final descent and landing. The high speeds at which lifting body aircraft operate make them dangerous to land. The parafoil would have been used to slow the vehicle and make landing safer. The landing gear consisted of skids rather than wheels: the skids worked like sleds so the vehicle would have slid to a stop on the ground.

Both the shape and size of the X-38 were different from that of the Space Shuttle. The Crew Return Vehicle would have fit into the payload bay of the shuttle. This does not, however, mean that it would have been small. The X-38 weighed 10,660 kg and was 9.1 meters long. The battery system, lasting nine hours, was to be used for power and life support. If the Crew Return Vehicle was needed, it would only take two to three hours for it to reach Earth.

The parafoil parachute, employed for landing, was derived from technology developed by the U.S. Army. This massive parafoil deploys in 5 stages for optimum performance. A drag chute would have been released from the rear of the X-38. This drag chute would have been used to stabilize and slow the vehicle down. The giant parafoil — area of 687 square meters — was then released. It would open in five steps (a process called staging). While the staging process only takes 45 seconds, it is important for a successful chute deployment. Staging prevents high-speed winds from tearing the parafoil.

The spacecraft's landing was to be completely automated. Mission Control would have sent coordinates to the onboard computer system. This system would also have used wind sensors and the Global Positioning System (a satellite-based coordinate system) to coordinate a safe trip home. Since the Crew Return Vehicle was designed with medical emergencies in mind, it made sense that the vehicle could find its way home automatically in the event that crew members were incapacitated or injured. If there was a need, the crew would have the capability to operate the vehicle by switching to the backup systems. In addition, seven high altitude low opening (HALO) parachute packs were included in the crew cabin, a measure designed to provide for the ability to bail out of the craft.

An Advanced Docking Berthing System (ADBS) was designed for the X-38 and the work on it led to the Low Impact Docking System the Johnson Space Center later created for the planned vehicles in Project Constellation.

The X-38 rescue vehicle was also known as the X-35 (but that designation was already allocated by the USAF to another vehicle) and X-CRV (experimental - Crew Return Vehicle)

DARPA

The Defense Advanced Research Projects Agency (DARPA) is an agency of the United States Department of Defense responsible for the development of new technology for use by the military. DARPA has been responsible for funding the development of many technologies which have had a major effect on the world, including computer networking, as well as NLS, which was both the first hypertext system, and an important precursor to the contemporary ubiquitous graphical user interface.

Its original name was simply Advanced Research Projects Agency (ARPA), but it was renamed to "DARPA" (for Defense) in March 1972, then renamed "ARPA" again in February 1993, and then renamed "DARPA" again in March 1996.

DARPA was established during 1958 (as ARPA) in response to the Soviet launching of Sputnik during 1957, with the mission of keeping U.S. military technology more sophisticated than that of the nation's potential enemies. From DARPA's own introduction.

DARPA's original mission, established in 1958, was to prevent technological surprise like the launch of Sputnik, which signaled that the Soviets had beaten the U.S. into space. The mission statement has evolved over time.

Today, DARPA's mission is still to prevent technological surprise to the US, but also to create technological surprise for our enemies.

DARPA is independent from other more conventional military R&D and reports directly to senior Department of Defense management. DARPA has around 240 personnel (about 140 technical) directly managing a $3.2 billion budget. These figures are "on average" since DARPA focuses on short-term (two to four-year) projects run by small, purpose-built teams.

Air Force Research Laboratory

The Air Force Research Laboratory (AFRL) is a scientific research organization operated by the United States Air Force Materiel Command dedicated to leading the discovery, development, and integration of affordable aerospace warfighting technologies; planning and executing the Air Force science and technology program; and provide warfighting capabilities to United States air, space, and cyberspace forces. It controls the entire Air Force science and technology research budget which was $2.4 billion in 2006.

The Laboratory was formed at Wright-Patterson Air Force Base, Ohio on 31 October 1997 as a consolidation of four Air Force laboratory facilities (Wright, Phillips, Rome, and Armstrong) and the Air Force Office of Scientific Research under a unified command. The Laboratory is composed of 8 technical directorates, 1 wing, and the Office of Scientific Research. Each technical directorate emphasizes a particular area of research within the AFRL mission which it specializes in performing experiments in conjunction with universities and contractors. Since the Laboratory's formation in 1997, it has conducted numerous experiments and technical demonstrators in conjunction with NASA, Department of Energy National Laboratories, DARPA, and other research organizations within the Department of Defense. Notable projects include the X-37, X-40, X-53, HTV-3X, YAL-1A, Advanced Tactical Laser, and the Tactical Satellite Program.

The Laboratory may face problems in the future as 40 percent of its workers are slated to retire over the next two decades while since 1980 the United States has not produced enough science and engineering degrees to keep up with demand.

NASA X-43

The X-43 is an unmanned experimental hypersonic aircraft with multiple planned scale variations meant to test various aspects of hypersonic flight. It was part of NASA's Hyper-X program. It has set several airspeed records for jet-propelled aircraft. A winged booster rocket with the X-43 itself at the tip, called a "stack", is launched from a carrier plane. After the booster rocket (a modified first stage of the Pegasus rocket) brings the stack to the target speed and altitude, it is discarded, and the X-43 flies free using its own engine, a scramjet.

The initial version, the X-43A, was designed to operate at speeds greater than Mach 7 (4,700 mph; 7,600 km/h) at altitudes of 30,000 m or more. The X-43A is a single-use vehicle and is designed to crash into the ocean without recovery. Three of them have been built: the first was destroyed; the other two have successfully flown, with the scramjet operating for approximately 10 seconds, followed by a 10 minute glide and intentional crash.

The first flight in June 2001 failed when the stack spun out of control about 11 seconds after the drop from the B-52 carrier plane. It was destroyed by the Range Safety Officer, and it crashed into the Pacific Ocean. NASA attributed the crash to several inaccuracies in data modeling for this test, which led to an inadequate control system for the particular Pegasus rocket used.

The X-43A's successful second flight made it the fastest free flying air-breathing aircraft in the world.

The third flight of the X-43A set a new speed record of 12,144 km/h (7,546 mph), or Mach 9.8, on November 16, 2004. It was boosted by a modified Pegasus rocket which was launched from a B-52 at 13,157 meters (43,166 ft). After a free flight where the scramjet operated for about 10 seconds, the craft made a planned crash into the Pacific Ocean off the coast of southern California.

The most recent success in the X-plane series of aircraft until it was replaced by the X-51, the X-43 was part of NASA's Hyper-X program, involving the American space agency and contractors such as Boeing, Micro Craft Inc, Orbital Sciences Corporation and General Applied Science Laboratory (GASL). Micro Craft Inc. built the X-43A and GASL built its engine.

The Hyper-X Phase I is a NASA Aeronautics and Space Technology Enterprise program being conducted jointly by the Langley Research Center, Hampton, Virginia, and the Dryden Flight Research Center, Edwards, California. Langley is the lead center and is responsible for hypersonic technology development. Dryden is responsible for flight research.

Phase I was a seven-year, approximately $230 million, program to flight-validate scramjet propulsion, hypersonic aerodynamics and design methods.

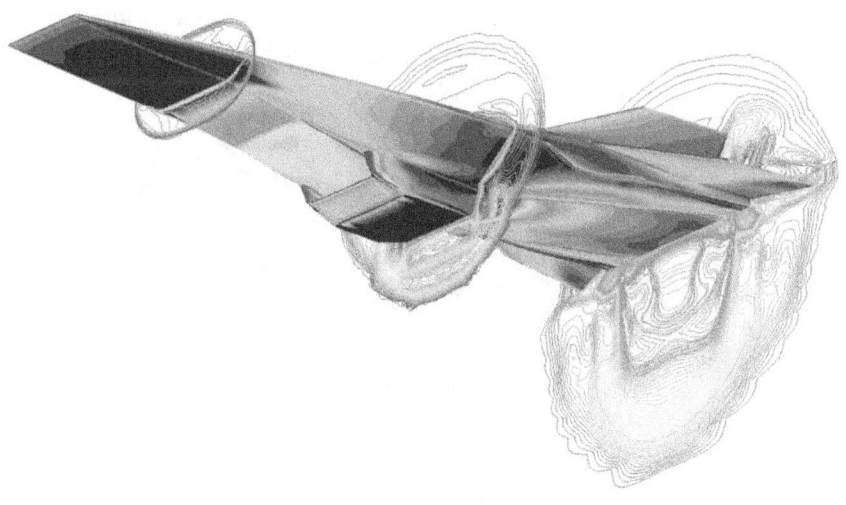

CFD image of the X-43A at Mach 7

Lockheed Martin X-44 MANTA

The Lockheed Martin X-44 MANTA (Multi-Axis No-Tail Aircraft) was a conceptual aircraft design by Lockheed Martin that has been studied by NASA and the U.S. Air Force. It was intended to test the feasibility of full yaw, pitch and roll control without tailplanes (horizontal or vertical). Attitude manipulation relies purely on 3D thrust vectoring. The aircraft design was derived from the F-22 Raptor and featured a stretched delta wing and no tail surfaces.

The X-44 was designed by Lockheed Martin to demonstrate the feasibility of an aircraft controlled by vectored thrust alone. The X-44 design had a reduced radar signature and was made more efficient by eliminating the tail and rudder surfaces, and instead using thrust vectors to provide yaw, pitch and roll control.

The X-44 MANTA design was based on the F-22, except without a tail and incorporated a full delta wing. The basic X-44 MANTA would entail a larger weapons payload and a greater fuel capacity than the F-22, due to its larger delta wing design. The MANTA was designed to have reduced mechanical complexity, increased fuel efficiency and greater agility. The X-44 MANTA combined the control and propulsion systems, using thrust vectoring. Funding for the X-44 program ended in 2000.

Boeing X-45

The Boeing X-45 unmanned combat air vehicle is a concept demonstrator for a next generation of completely autonomous military aircraft, developed by Boeing's Phantom Works. Manufactured by Boeing Integrated Defense Systems, the X-45 was a part of DARPA's J-UCAS project.

The larger X-45B design was modified to have even more fuel capacity and three times greater combat range, becoming the X-45C. Each wing's leading edge spans from the nose to the wingtip, giving the aircraft more wing area, and a planform very similar to the B-2 Spirits'. The first of the three planned X-45C

140

aircraft was originally scheduled to be completed in 2006, with capability demonstrations scheduled for early 2007. By 2010 Boeing hoped to complete an autonomous aerial refueling of the X-45C by a KC-135 Stratotanker. Boeing has displayed a mock-up of the X-45C on static displays at many airshows.

The X-45C portion of the program received $767 million from DARPA in October 2004, to construct and test three aircraft, along with several supplemental goals. The X-45C included an F404 engine. In July 2005 DARPA awarded an additional $175 million to continue the program, as well as implement autonomous Aerial Refueling technology.

As of March 2, 2006, the US Air Force has decided not to continue with the X-45 project. However, Boeing submitted a proposal to the Navy for a carrier based demonstrator version of the X-45, designated the X-45N.

Boeing X-46

The Boeing X-46 was a proposed unmanned combat air vehicle (UCAV) that was to be developed in conjunction with the U.S. Navy and DARPA as a naval carrier-based variant of the Boeing X-45 UCAV being developed for the U.S. Air Force. Two contracts for technology demonstrators were awarded in June 2000, to Boeing for the X-46A and to Northrop Grumman for the X-47A.

However, in April 2003, the Air Force and the Navy efforts were formally combined under the joint DARPA/USAF/Navy J-UCAV program, later renamed J-UCAS (Joint Unmanned Combat Air Systems), and the X-46 program was terminated as redundant. The J-UCAS program was later terminated.

A Navy-only N-UCAS demonstrator program started in the summer of 2006. Boeing will use material developed for the X-46 and X-45 to propose the X-45N as a naval UCAV demonstrator.

Northrop Grumman X-47A Pegasus

The Northrop Grumman X-47 is a demonstration Unmanned Combat Aerial Vehicle. The X-47 began as part of DARPA's J-UCAS program, and is now part of the United States Navy's UCAS-D program to create a carrier-based unmanned aircraft. Unlike the Boeing X-45, initial Pegasus development was company-funded. The original vehicle carries the designation X-47A Pegasus, while the follow-on naval version is designated X-47B.

The US Navy did not commit to practical UCAV efforts until mid-2000, when the service awarded contracts of US$2 million each to Boeing and Northrop Grumman for a 15-month concept-exploration program. Design considerations for a naval UCAV included dealing with the corrosive salt-water environment, deck handling for launch and recovery, integration with command and control systems, and operation in a carrier's high electromagnetic interference environment. The Navy was also interested in using their UCAVs for reconnaissance missions, penetrating protected airspace to identify targets for the attack waves.

Boeing X-48

The X-48 is an experimental unmanned aerial vehicle (UAV) for investigation into the characteristics of blended wing body (BWB) aircraft, a type of flying wing. It is currently under development by Boeing and NASA.

Boeing Phantom Works is focusing current research on a pair of models, called the X-48B, which were built under contract by Cranfield Aerospace in the United Kingdom. Norman Princen, Boeing's chief engineer for the project, said, "Earlier wind-tunnel testing and the upcoming flight testing are focused on learning more about the BWB's low-speed flight-control characteristics, especially during takeoffs and landings. Knowing how accurately our models predict these characteristics is an important step in the further development of this concept."

The X-48B has a 21-foot (6.4 m) wingspan, weighs 392-pound (178 kg), and is built from composite materials. It is powered by three small turbojet engines and is expected to fly at up to 120 kn (220 km/h) and reach an altitude of 10,000 feet (3,000 m). The X-48B is a scaled down from a conceptual 240-foot wide design. Though passenger versions of the X-48B have been proposed, the design has a higher probability  of first being used for a military transport.

Wind tunnel testing on a 12 ft wide blended wing body model was completed in September 2005. During April and May 2006, NASA performed wind tunnel tests on X-48B Ship 1, an 8.5% scale model, at a facility shared by Langley and Old Dominion University. After the wind tunnel testing, the vehicle was shipped to NASA's Dryden Flight Research Center at Edwards Air Force Base to serve as a backup to X-48B Ship 2 for flight testing. In November 2006, ground testing began at Dryden, to validate the aircraft's systems integrity, telemetry and communications links, flight-control software and taxi and takeoff characteristics.

Piasecki X-49

The Piasecki X-49 is a four-bladed, twin-engined, experimental compound helicopter under development by Piasecki Aircraft. The X-49A is

based on the airframe of a Sikorsky YSH-60F Seahawk, but utilizes Piasecki's proprietary vectored thrust ducted propeller (VTDP) design and includes the addition of lifting wings. The concept of the experimental program is to apply the VTDP technology to a production military helicopter to determine any benefit gained through increases in performance or useful load.

"SpeedHawk" is a concept aircraft based on applying X-49A compounding concepts to a production UH-60 Black Hawk offering better performance, range, and increases in useful load. The "SpeedHawk" aircraft includes an SPU (third engine), high forward-swept wing concept, a 45 inch cabin extending fuselage "plug", and several other drag reducing and performance-oriented improvements, including a rotorhead fairing, landing gear streamlining, and a fly-by-wire flight control system.

Boeing X-51

The Boeing X-51 (also known as X-51 WaveRider) is an unmanned scramjet demonstration aircraft for hypersonic (Mach 6, approximately 4,000 miles per hour (6,400 km/h) at altitude) flight testing. It successfully completed its first free-flight on 26 May 2010 and also achieved the longest duration flight at speeds over Mach 5.

The X-51 "WaveRider" program is run as a cooperative effort of the United States Air Force, DARPA, NASA, Boeing, and Pratt & Whitney Rocketdyne. The program is managed by the Propulsion Directorate within the

United States Air Force Research Laboratory (AFRL). The X-51 had its first captive flight attached to a B-52 in December 2009.

In the 1990s, the Air Force Research Laboratory (AFRL) began the HyTECH program for hypersonic propulsion. Pratt & Whitney received a contract from the AFRL to develop a hydrocarbon-fueled scramjet engine which led to the development of the SJX61 engine.

The SJX61 engine was originally meant for NASA's X-43C, which was eventually canceled. The engine was applied to the AFRL's Scramjet Engine Demonstrator program in late 2003. The scramjet flight test vehicle was designated X-51 on 27 September 2005.

In flight demonstrations, the X-51 is carried by a B-52 to an altitude of about 50,000 feet (15.2 kilometers) and then released over the Pacific Ocean.

The X-51 is initially propelled by an MGM-140 ATACMS solid rocket booster to approximately Mach 4.5, before it is jettisoned.

Then the vehicle's Pratt & Whitney Rocketdyne SJY61 scramjet takes over and accelerates it to a top flight speed near Mach 6.

The X-51 uses JP-7 fuel for the SJY61 scramjet, carrying some 270 lb (120 kg) onboard.

Boeing X-53 Active Aeroelastic Wing

The X-53 Active Aeroelastic Wing (AAW) development program is a completed research project that was undertaken jointly by the Air Force Research Laboratory (AFRL), Boeing Phantom Works and NASA's Dryden Flight Research Center, where the technology was flight tested on a modified McDonnell Douglas F/A-18 Hornet. Active Aeroelastic Wing Technology is a technology that integrates wing aerodynamics, controls, and structure to harness and control wing aeroelastic twist at high speeds and dynamic pressures. By using multiple leading and trailing edge controls like "aerodynamic tabs", subtle amounts of aeroelastic twist can be controlled to provide large amounts of wing control power, while minimizing maneuver air loads at high wing strain conditions or aerodynamic drag at low wing strain conditions.

Gerry Miller and Jan Tulinius led the development of the initial concept during wind tunnel testing in the mid 1980s under Air Force contract. The designation "X-52" was skipped in sequence to avoid confusion with the B-52 Stratofortress bomber.

Gulfstream X-54

The Gulfstream X-54 is a research and demonstration aircraft, under development in the United States by Gulfstream Aerospace, that is planned for use in sonic boom and supersonic transport research.

Initiated during 2008, the X-54 project is intended to produce an experimental aircraft capable of supersonic speeds. The X-54A is intended to produce test data on sonic boom effects in support of future supersonic transport design and regulation. Current regulations prohibit supersonic flight over land areas in the United States; the X-54 is part of Gulfstream's efforts to have the regulations altered to allow for supersonic transports to be commercially viable.

The X-54A is being developed by Gulfstream Aerospace and is intended to be powered by two Rolls-Royce Tay turbofan engines. Although the aircraft has received an 'X' series designation in the U.S. Department of Defense's Mission Designation System at the request of NASA, neither the U.S. military nor NASA is currently involved in the project.

Although Gulfstream has made little comment about the X-54A project, at the 2008 National Business Aviation Association convention a Gulfstream executive stated that Gulfstream's work on advanced technologies for supersonic flight had been ongoing "for some time" and that a "complete airplane designed for low [sonic] boom" would possibly "have X-54 painted on the side of it."

The X-54A may be connected to Gulfstream's "Sonic Whisper" program, trademarked in 2005 as an aircraft design to "reduce boom intensities during supersonic flight." Some sources claim that the X-54A is based on the Lockheed F-104 Starfighter; this conflicts with the description of the aircraft by the DOD.

Lockheed Martin X-55

The Lockheed Martin X-55 Advanced Composite Cargo Aircraft (ACCA) is an experimental twin jet engined transport aircraft. It is intended to demonstrate new cargo-carrier capabilities using advanced composites. A project of the United States Air Force's Air Force Research Laboratory, it was built by the international aerospace company Lockheed Martin, at its Advanced Development Programs (Skunk Works) facility in Palmdale, California.

Boeing Bird of Prey

The Boeing Bird of Prey was a black project aircraft, intended to demonstrate stealth technology. It was developed by McDonnell Douglas and Boeing in the 1990s. Funded by the company at a price of $67 million, it was a low cost program compared to many other programs of similar scale. It developed technology and materials which would later be used on Boeing's X-45 unmanned combat air vehicle. As an internal project, this aircraft was not given an X-plane designation. There are no public plans to make this a production aircraft. It is characterized as a technology demonstrator.

DARPA Falcon Project

The DARPA Falcon Project (Force Application and Launch from CONtinental United States) is a two-part joint project between the Defense Advanced Research Projects Agency (DARPA) and the United States Air Force (USAF) and is part of Prompt Global Strike. One part of the program aims to develop a reusable, rapid-strike Hypersonic Weapon System (HWS), now retitled the Hypersonic Cruise Vehicle (HCV), and the other is for the development of a launch system capable of accelerating a HCV to cruise speeds, as well as launching small satellites into earth orbit. This two-part program was announced in 2003 and continued into 2006.

The latest project to be announced under the Falcon banner was a fighter-sized unmanned aircraft called Blackswift which would take off from a runway and accelerate to Mach 6 before completing its mission and landing again. The memo of understanding between DARPA and the USAF on Blackswift—also known as the HTV-3X—was signed in September 2007. The Blackswift HTV-3X did not receive needed funding and was canceled in October 2008.

NASA Pathfinder

The NASA Pathfinder, NASA Pathfinder Plus, NASA Centurion and NASA Helios Prototype were an evolutionary series of solar- and fuel-cell-system-powered unmanned aerial vehicles. AeroVironment, Inc. developed the vehicles under NASA's Environmental Research Aircraft and Sensor Technology (ERAST) program. They were built to develop the technologies that would allow long-term, high-altitude aircraft to serve as "atmospheric satellites", to perform atmospheric research tasks as well as serve as communications platforms.

AeroVironment initiated its development of full-scale solar-powered aircraft with the Gossamer Penguin and Solar Challenger vehicles in the late 1970s and early 1980s, following the pioneering work of Robert Boucher, who built the first solar-powered flying models in 1974. Under ERAST, AeroVironment built four generations of long endurance unmanned aerial vehicles (UAVs), the first of which was the Pathfinder.

Pathfinder-Plus

During 1998, the Pathfinder was modified into the longer-winged Pathfinder-Plus configuration. It used four of the five sections from the original Pathfinder wing, but substituted a new 44 feet (13 m) long center wing section that incorporated a high-altitude airfoil designed for the follow-on Centurion/Helios. The new section was twice as long as the original, and increased the overall wingspan of the craft from 98.4 feet (30.0 m) to 121 feet (37 m).

The new center section was topped by more-efficient silicon solar cells developed by SunPower Corporation of Sunnyvale, California, which could convert almost 19 percent of the solar energy they receive to useful electrical energy to power the craft's motors, avionics and communication systems. That compared with about 14 percent efficiency for the older solar arrays that cover most of the surface of the mid- and outer wing panels from the original Pathfinder. Maximum potential power was boosted from about 7,500 watts on Pathfinder to about 12,500 watts on Pathfinder-Plus. The number of electric motors was increased to eight, and the motors used were more powerful units, designed for the follow-on aircraft.

The Pathfinder-Plus development flights flown at PMRF in the summer of 1998 validated power, aerodynamic, and systems technologies for its successor, the Centurion. On August 6, 1998, Pathfinder-Plus proved its design by raising the national altitude record to 80,201 feet (24,445 m) for solar-powered and propeller-driven aircraft.

Rutan Boomerang

The Rutan Model 202 Boomerang is an aircraft designed and built by Burt Rutan. The design was intended to be a multi-engine aircraft that would not become dangerously difficult to control in the event of failure of a single engine. The result is an aircraft with a very asymmetrical appearance.

In 1997, avionics entrepreneur Ray Morrow and his son, Neil Morrow, founded an air taxi company. They settled on a modified version of Rutan's Boomerang design, which they designated the MB-300. They determined that the best business approach would be to manufacture the aircraft and run the air taxi services themselves. So Ray Morrow founded Morrow Aircraft Corporation in order to design and manufacture the MB-300. In the meantime, they started the SkyTaxi company using Cessna 414s as interim aircraft. In 1999, Morrow Aircraft Corporation applied to the Federal Aviation Administration (FAA) of the United States for a type certificate for the MB-300. In 2000, the FAA published a notice seeking comments on Morrow Aircraft's proposal to use an electronic engine control system (FADEC) in place of the engine's mechanical system.

Scaled Composites Proteus

The Scaled Composites Model 281 Proteus is a tandem-wing high-endurance aircraft designed by Burt Rutan to investigate the use of aircraft as high altitude telecommunications relays. The Proteus is actually a multi-mission vehicle, able to carry various payloads on a ventral pylon. An extremely efficient design, the Proteus can orbit a point at over 65,000 feet (19,800 m) for more than 18 hours. It is currently owned by Northrop Grumman.

BAE Systems Mantis

The BAE Systems Mantis Unmanned Autonomous System Advanced Concept Technology Demonstrator is a British demonstrator programme for Unmanned Combat Air Vehicle (UCAV) technology. It is the world's first unmanned autonomous aircraft. The Mantis is a twin-engined turboprop-powered UCAV with a wingspan of approximately 22 m, broadly comparable to the MQ-9 Reaper.

Other partners involved in Phase 1 of the Mantis programme include the UK Ministry of Defence, Rolls-Royce, QinetiQ, GE Aviation, L3 Wescam, Meggitt and Lola.

Dassault nEUROn

The Dassault nEUROn is an experimental Unmanned Combat Air Vehicle (UCAV) being developed with international cooperation, led by the French company Dassault Aviation.

As a UCAV, nEUROn will be significantly larger and more advanced than other well-known UAV systems like the MQ-1 Predator, with ranges, payloads and capabilities that approach those of manned fighter aircraft. Although the project is not yet closely defined, illustrations and statements by the consortium partners indicate that the nEUROn is envisioned as a competitive system with the American J-UCAS program's Boeing X-45C or Northrop-Grumman X-47B.

Indeed, Saab's February 9, 2006 release notes that nEUROn will be a demonstrator measuring 10 m long by 12 m wide and weighing in at 5 tons. This is roughly the size of a Mirage 2000 fighter. The aircraft will have unmanned autonomous air-to-ground attack capabilities with precision guided munitions, relying on an advanced stealth airframe design to penetrate undetected. Another feature being contemplated is the ability to control squad flight in automatic mode from an advanced fighter like the Rafale or JAS 39 Gripen platform, grouping the nEUROns and controlling the group in a manner similar to many combat real-time strategy computer games.

Payen Pa 49

The Payen Pa 49 Katy was a small experimental French turbojet powered tailless aircraft, first flown in 1954, was the first French aircraft of this kind and the smallest jet aircraft of its day.

The first flight of what was now the Pa 49A took place on 22 January 1954 at Melun-Villaroche flown by Tony Ochsenbein, a comparatively inexperienced pilot, who had previously logged only 30 minutes on jets. Ten hours of manufacturer's testing was followed, in April 1954, by assessment at the Centre d'Essais en Vol (CEV), Brétigny-sur-Orge.

The aerobatic ability of the Pa 49 was established. At the CEV it was fitted with a split rudder airbrake; the two surfaces of the rudder separated from just below the tip, driven via faired external links near the bottom, into a V at the hinge for braking, rotating together for yaw control.

This airbrake was designed by Fléchair SA, a company founded by Payen. At the time of its appearance at the 12th Salon International d'Aeronautique at Paris, in 1957, the undercarriage legs were faired and the main wheels enclosed in spats and the aircraft renamed the Pa 49B.

For a time the nosewheel was also spatted. There were plans for a version with a retractable undercarriage, but this did not come about.

Junkers Ju 287

The Junkers Ju 287 was a German flying testbed built to develop the technology required for a multi-engine jet bomber. It was powered by four Junkers Jumo 004 engines, featured a revolutionary forward-swept wing and was built largely from scavenged components from other aircraft.

The flying prototype and an unfinished third prototype were captured by the Red Army in the closing stages of World War II and the design was further developed in the Soviet Union after the end of the war.

VFW VAK 191B

The VFW VAK 191B was an experimental German VTOL nuclear strike fighter of the early 1970s. Designed and built by the Vereinigte Flugtechnische Werke (VFW) it was intended to lead to a replacement for the Fiat G.91.

The VAK 191B was produced by the German company Vereinigte Flugtechnische Werke (VFW). Initially, Fiat of Italy was also involved but dropped out in 1967, though it remained as a major sub-contractor. VAK was the abbreviation for Vertikalstartendes Aufklärungs- und Kampfflugzeug (V/STOL Reconnaissance and Strike Aircraft).

EADS Barracuda

The EADS Barracuda is a European unmanned aerial vehicle (UAV) currently under development by EADS, intended for the role of reconnaissance and also combat (UCAV). The aircraft is a joint venture between Germany and Spain.

Development of the project was stopped after the first prototype crashed at sea while approaching for landing during a test flight. The program was resumed in 2008, with a second prototype being completed in November 2008. The rebuilt Barracuda underwent a series of successful flight tests in Goose Bay, Canada during July 2009.

The Barracuda is primarily in competition with the Dassault nEUROn for strategic and defensive contracts. Both are stealthy and have a maximum air speed of around Mach 0.85 . While Germany and Spain are behind the Barracuda, France, Italy and Sweden, among others, are funding the nEUROn. Not much is known about the Barracuda as it is still in development. However the Barracuda is thought to have an operating ceiling of around 20,000 ft and carries a maximum payload of 300 kg.

The Barracuda originated as a UAV design study, intended to push EADS into the market for medium-altitude long-range UAVs, a market they view as dominated by the United States and Israel. Its official debut was at the 2006

International Aerospace Exhibition, where military applications and specifications for the Barracuda were revealed. EADS' current focus is to get the Barracuda certified for unregulated flight in Germany's designated airspace, while the long-term goal is to have it certified for non-segregated airspace.

Solar Impulse

Solar Impulse is a Swiss long-range solar powered aircraft project being undertaken at the École Polytechnique Fédérale de Lausanne. The project eventually hopes to achieve the first circumnavigation of the Earth by a piloted fixed-wing aircraft using only solar power. The project is led by Swiss psychiatrist and aeronaut Bertrand Piccard, who co-piloted the first balloon to circle the world non-stop, and Swiss businessman André Borschberg.

The first aircraft, bearing the Swiss aircraft registration code of HB-SIA, is a single-seater monoplane, capable of taking off under its own power, and intended to remain airborne up to 36 hours. This aircraft first flew an entire diurnal solar cycle, including nearly 9 hours of night flying, in a 26-hour flight on 7–8 July 2010. Building on the experience of this prototype, a slightly larger follow-on design (HB-SIB) is planned to make circumnavigation of the globe in 20–25 days; this flight was planned for 2014 but following a structural failure of the main spar, a more likely date is 2015.

Piccard initiated the Solar Impulse project in 2003. By 2009, he had assembled a multi-disciplinary team of 50 specialists from six countries, assisted by about 100 outside advisers. The project is financed by a number of private companies. The four main partners are Deutsche Bank, Omega SA, Solvay, and

157

Schindler. Other partners include Bayer MaterialScience, Altran and Swisscom. Other supporters include Clarins, Semper, Toyota, BKW and STG. The EPFL, the European Space Agency (ESA) and Dassault have provided additional technical expertise.

Conceptual drawing of a supersonic biplane

(Credit: Christine Daniloff/MIT News based on an original drawing courtesy of Obayashi laboratory, Tohoku University)

From: *http://www.sciencedaily.com/releases/2012/03/120319163811.htm*

Double bubble

MIT's D "double bubble" series design concept is based on a modified "tube-and-wing" structure that has a very wide fuselage to provide extra lift.

The aircraft would be used for domestic flights to carry 180 passengers in a coach cabin roomier than that of a Boeing 737-800. (Credit: MIT/Aurora Flight Sciences)

From:

http://www.sciencedaily.com/releases/2010/05/100517162834.htm

Box wing design

Lockheed Martin took an entirely different approach. Its engineers proposed a box wing design, in which a front wing mounted on the lower belly of the plane is joined at the tips to an aft wing mounted on top of the plane.

The company has studied the box wing concept for three decades, but has been waiting for lightweight composite materials, landing gear technologies, hybrid laminar flow and other tools to make it a viable configuration.

Lockheed's proposal combines the unique design with a Rolls Royce Liberty Works Ultra Fan Engine.

This engine has a bypass ratio that is approximately five times greater than current engines, pushing the limits of turbofan technology.

From:

http://www.nasa.gov/topics/aeronautics/features/greener_aircraft.html

Another Take on Supersonic

Our ability to fly at supersonic speeds over land in civil aircraft depends on our ability to reduce the level of sonic booms.

NASA has been exploring a variety of options for quieting the boom, starting with design concepts and moving through wind tunnel tests to flight tests of new technologies.

This rendering of a possible future civil supersonic transport shows a vehicle that is shaped to reduce the sonic shockwave signature and also to reduce drag.

From:

Flying Wing a Regular Sight

This computer-generated image shows a possible future "flying wing" aircraft, very efficiently and quietly in flight over populated areas.

This kind of design, produced by Northrop Grumman, would most likely carry cargo at first and then also carry passengers.

This design is among those presented to NASA at the end of 2011 by companies that conducted NASA-funded studies into aircraft that could enter service in 2025.

From:

*http://www.nasa.gov/images/content/619970main_passenger_original_4x3_f
ull.jpg*

AMELIA Climbs High

This computer rendering shows AMELIA (Advanced Model for Extreme Lift and Improved Aeroacoustics), a possible future hybrid wing body-type subsonic vehicle with short takeoff and landing capabilities.

Produced through a three-year NASA Research Announcement grant with the California Polytechnic State Institute, AMELIA's ability for steeper ascents and descents could reduce community noise levels on takeoff and landing.

A model of this configuration is scheduled for testing in a NASA wind tunnel in the fall of 2011.

From:

*http://www.nasa.gov/images/content/575317main_amelia_5400x3393_full.jp
g*

Boxed-Wing Reduces Drag

This artist's concept shows a possible future subsonic aircraft using a boxed- or joined-wing configuration to reduce drag and increase fuel efficiency.

This design of an aircraft that could enter service in the 2020 timeframe is one of a number of designs being explored by NASA with teams of researchers from industry and universities.

This artist's concept shows a possible future subsonic aircraft using a boxed- or joined-wing configuration to reduce drag and increase fuel efficiency.

This design of an aircraft that could enter service in the 2020 timeframe is one of a number of designs being explored by NASA with teams of researchers from industry and universities.

From:

http://www.nasa.gov/images/content/575281main_lockheed_avc_1245x910_ full.jpg

Down to Earth Future Aircraft

The Subsonic Ultra Green Aircraft Research, or SUGAR, Volt future aircraft design comes from the research team led by The Boeing Company.

The Volt is a twin-engine concept with a hybrid propulsion system that combines gas turbine and battery technology, a tube-shaped body and a truss-braced wing mounted to the top of the aircraft.

This aircraft is designed to fly at Mach 0.79 carrying 154 passengers 3,500 nautical miles.

The SUGAR Volt is among the designs presented in April 2010 to the NASA Aeronautics Research Mission Directorate for its NASA Research Announcement-funded studies into advanced aircraft that could enter service in the 2030-2035 timeframe.

From:

Green Supersonic Machine

This future aircraft design concept for supersonic flight over land comes from the team led by the Lockheed Martin Corporation.

The team used simulation tools to show it was possible to achieve over-land flight by dramatically lowering the level of sonic booms through the use of an "inverted-V" engine-under wing configuration. Other revolutionary technologies help achieve range, payload and environmental goals.

This concept is one of two designs presented in April 2010 to the NASA Aeronautics Research Mission Directorate for its NASA Research Announcement-funded studies into advanced supersonic cruise aircraft that could enter service in the 2030-2035 timeframe.

From:

http://www.nasa.gov/images/content/453787main_lm_supers_upper_engine _original_full.jpg

An Iconic Idea

The "Icon-II" future aircraft design concept for supersonic flight over land comes from the team led by The Boeing Company.

A design that achieves fuel burn reduction and airport noise goals, it also achieves large reductions in sonic boom noise levels that will meet the target level required to make supersonic flight over land possible.

This concept is one of two designs presented in April 2010 to the NASA Aeronautics Research Mission Directorate for its NASA Research Announcement-funded studies into advanced supersonic cruise aircraft that could enter service in the 2030-2035 timeframe.

From:

http://www.nasa.gov/images/content/453799main_boeing_supers_original_f ull.jpg

NEW AIRCRAFT

ION THRUSTER

About the ion thruster

Speaking about a new ionic engine means to speak about a new aircraft.

The chapter presents shortly the actual ionic engines (called ion thrusters) and the new ionic (pulse) engines proposed by the author.

Ionic engine (ion thruster, which accelerates the positive ions through a potential difference) is about 10 times more effective than classic system based on combustion.

We can still improve the efficiency of 10-50 times if one uses pulses of positive ions accelerated in a cyclotron mounted on the ship; the efficiency can easily grow for 1000 times if the positive ions will be accelerated in a high energy synchrotron, synchrocyclotron or isochronous cyclotron (1-100 GeV). In this, the big classic synchrotron is reduced to a ring surface (magnetic core).

Future (ionic) engine will have mandatory a circular particle accelerator (high or very high energy).

We can thus increase the speed and autonomy of the ship using a less quantity of fuel and power.

One can use synchrotron radiation (synchrotron light, high intensity beams), like high intensity (X-ray or Gamma ray) radiation, as well. In this case will be a beam engine (not an ionic engine), it'll use only the power (energy, which can be solar energy, nuclear energy, or both) and so we will remove the fuel.

It proposes using a powerful LINAC at the exit of synchrotron (especially when one accelerates electrons) to not lose energy by photons premature emission.

With a new ionic engine one builds a new aircraft, which can travel through water and. This new aircraft will can accelerate directly, without an additional combustion engine and without gravity assists from other planets.

An *ion thruster* is a form of electric propulsion used for spacecraft propulsion that creates thrust by accelerating ions. Ion thrusters are characterized by how they accelerate the ions, using either electrostatic or electromagnetic force.

Electrostatic ion thrusters use the Coulomb Force and accelerate the ions in the direction of the electric field. Electromagnetic ion thrusters use the Lorentz Force to accelerate the ions. Note that the term "ion thruster" frequently denotes the electrostatic or gridded ion thrusters, only.

The thrust created in ion thrusters is very small compared to conventional chemical rockets, but a very high specific impulse, or propellant efficiency, is obtained.

Due to their relatively high power needs, given the specific power of power supplies, and the requirement of an environment void of other ionized particles, ion thrust propulsion currently is only practicable in outer space.

The first experiments with ion thrusters were carried out by Robert Goddard at Clark College from 1916-1917. The technique was recommended for near-vacuum conditions at high altitude, but thrust was demonstrated with ionized air streams at atmospheric pressure. The idea appeared again in Hermann Oberth's "Wege zur Raumschiffahrt" (Ways to Spaceflight), published in 1923.

A working ion thruster was built by Harold R. Kaufman in 1959 at the NASA Glenn facilities. It was similar to the general design of a gridded electrostatic ion thruster with mercury as its fuel. Suborbital tests of the engine followed during the 1960s and in 1964 the engine was sent into a suborbital flight aboard the Space Electric Rocket Test 1 (SERT 1). It successfully operated for the planned 31 minutes before falling back to Earth.

Hall effect thruster

The Hall effect thruster was studied independently in the U.S. and the USSR in the 1950s and 60s. However, the concept of a Hall thruster was only developed into an efficient propulsion device in the former Soviet Union, whereas in the U.S., scientists focused instead on developing gridded ion thrusters.

Hall effect thrusters were operated on Soviet satellites since 1972. Until the 1990s they were mainly used for satellite stabilization in North-South and in East-West directions. Some 100-200 engines completed their mission on Soviet and Russian satellites until the late 1990s. Soviet thruster design was introduced to the West in 1992 after a team of electric propulsion specialists, under the support of the Ballistic Missile Defense Organization, visited Soviet laboratories.

Ion thrusters utilize beams of ions (electrically charged atoms or molecules) to create thrust in accordance with Newton's third law. The method of accelerating the ions varies, but all designs take advantage of the charge/mass ratio of the ions. This ratio means that relatively small potential differences can create very high exhaust velocities. This reduces the amount of reaction mass or fuel required, but increases the amount of specific power required compared to chemical rockets. Ion thrusters are therefore able to achieve extremely high specific impulses. The drawback of the low thrust is low spacecraft acceleration because the mass of current electric power units is directly correlated with the amount of power given. This low thrust makes ion thrusters unsuited for launching spacecraft into orbit, but they are ideal for in-space propulsion applications.

Hall effect thrusters accelerate ions with the use of an electric potential maintained between a cylindrical anode and a negatively charged plasma which forms the cathode. The bulk of the propellant (typically xenon or bismuth gas) is introduced near the anode, where it becomes ionized, and the ions are attracted towards the cathode, they accelerate towards and through it, picking up electrons as they leave to neutralize the beam and leave the thruster at high velocity.

The anode is at one end of a cylindrical tube, and in the center is a spike which is wound to produce a radial magnetic field between it and the surrounding tube. The ions are largely unaffected by the magnetic field, since they are too massive. However, the electrons produced near the end of the spike to create the cathode are far more affected and are trapped by the magnetic field, and held in place by their attraction to the anode. Some of the electrons spiral down towards the anode, circulating around the spike in a Hall current. When they reach the anode they impact the uncharged propellant and cause it to be ionized, before finally reaching the anode and closing the circuit.

Gridded electrostatic ion thrusters

Gridded electrostatic ion thrusters commonly utilize xenon gas. This gas has no charge and is ionized by bombarding it with energetic electrons. These electrons can be provided from a hot cathode filament and accelerated in the electrical field of the cathode fall to the anode (Kaufman type ion thruster). Alternatively, the electrons can be accelerated by the oscillating electric field induced by an alternating magnetic field of a coil, which results in a self-sustaining discharge and omits any cathode (radiofrequency ion thruster).

The positively charged ions are extracted by an extraction system consisting of 2 or 3 multi-aperture grids. After entering the grid system via the plasma sheath the ions are accelerated due to the potential difference between the first and second grid (named screen and accelerator grid) to the final ion energy of typically 1-2 keV, thereby generating the thrust.

Ion thrusters emit a beam of positive charged xenon ions only. In order to avoid the charging-up of the spacecraft another cathode, placed near the engine, emits additional electrons (basically the electron current is the same as the ion current) into the ion beam. This also prevents the beam of ions from returning to the spacecraft and thereby cancelling the thrust.

Gridded electrostatic ion thruster research (past/present):

NASA Solar electric propulsion Technology Application Readiness (NSTAR)

NASA's Evolutionary Xenon Thruster (NEXT)

Nuclear Electric Xenon Ion System (NEXIS)

High Power Electric Propulsion (HiPEP)

EADS Radio-Frequency Ion Thruster (RIT)

Dual-Stage 4-Grid (DS4G)

Field Emission Electric Propulsion

Field Emission Electric Propulsion (FEEP) thrusters use a very simple system of accelerating liquid metal ions to create thrust. Most designs use either cesium or indium as the propellant. The design consists of a small propellant reservoir that stores the liquid metal, a very small slit that the liquid flows through, and then the accelerator ring.

Cesium and indium are used due to their high atomic weights, low ionization potentials, and low melting points. Once the liquid metal reaches the inside of the slit in the emitter, an electric field applied between the emitter and the accelerator ring causes the liquid metal to become unstable and ionize.

This creates a positive ion, which can then be accelerated in the electric field created by the emitter and the accelerator ring. These positively charged ions are then neutralized by an external source of electrons in order to prevent charging of the spacecraft hull.

Pulsed Inductive Thrusters

Pulsed Inductive Thrusters (PIT) use pulses of thrust instead of one continuous thrust, and have the ability to run on power levels in the order of Megawatts (MW).

PITs consist of a large coil encircling a cone shaped tube that emits the propellant gas. Ammonia is the gas commonly used in PIT engines.

For each pulse of thrust the PIT gives, a large charge first builds up in a group of capacitors behind the coil and is then released. This creates a current that moves circularly. The current then creates a magnetic field in the outward radial direction (Br), which then creates a current in the ammonia gas that has just been released in the opposite direction of the original current.

This opposite current ionizes the ammonia and these positively charged ions are accelerated away from the PIT engine due to the electric field crossing with the magnetic field Br, which is due to the Lorentz Force.

Magnetoplasmadynamic

Magnetoplasmadynamic (MPD) thrusters and Lithium Lorentz Force Accelerator (LiLFA) thrusters use roughly the same idea with the LiLFA thruster building off of the MPD thruster.

Hydrogen, argon, ammonia, and nitrogen gas can be used as propellant. The gas first enters the main chamber where it is ionized into plasma by the electric field between the anode and the cathode. This plasma then conducts electricity between the anode and the cathode.

This new current creates a magnetic field around the cathode which crosses with the electric field, thereby accelerating the plasma due to the Lorentz Force. The LiLFA thruster uses the same general idea as the MPD thruster, except for two main differences.

The first difference is that the LiLFA uses lithium vapor, which has the advantage of being able to be stored as a solid.

The other difference is that the cathode is replaced by multiple smaller cathode rods packed into a hollow cathode tube.

The cathode in the MPD thruster is easily corroded due to constant contact with the plasma. In the LiLFA thruster the lithium vapor is injected into the hollow cathode and is not ionized to its plasma form/corrode the cathode rods until it exits the tube.

The plasma is then accelerated using the same Lorentz Force.

Electrodeless Plasma Thrusters

Electrodeless Plasma Thrusters have two unique features, the removal of the anode and cathode electrodes and the ability to throttle the engine.

The removal of the electrodes takes away the factor of erosion which limits lifetime on other ion engines. Neutral gas is first ionized by electromagnetic waves and then transferred to another chamber where it is accelerated by an oscillating electric and magnetic field, also known as the ponderomotive force.

This separation of the ionization and acceleration stage give at the engine the ability to throttle the speed of propellant flow, which then changes the thrust magnitude and specific impulse values [1].

Plasma Micro Thruster

In the picture number 1 one presents „A Plasma Micro Thruster" Schematic and Prototype (see Figure 1, and [2]).

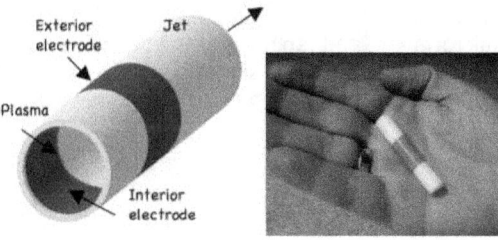

Plasma Micro Thruster Schematic and Prototype

Exterior electrode

Jet

Plasma

Interior electrode

Plasma Micro Thruster Jet Output

Fig. 1: *Plasma Micro Thruster, Schematic and Prototype*

THE HIPEP ENGINE

Powerful ion engine relies on microwaves

A powerful new ion propulsion system has been successfully ground-tested by NASA. The High Power Electric Propulsion ion engine trial marks the "first measurable milestone" for the ambitious $3 billion Project Prometheus, says director Alan Newhouse.

The HiPEP engine is the first tested propulsion technology with the potential power and longevity to thrust spacecraft as far as Jupiter without gravity assists from other planets.

These assists involve slingshot maneuvers around planets and can boost the speed of craft significantly. But they require specific planetary alignments, meaning suitable launch dates are rare.

In contrast, a probe powered by a HiPEP engine could launch any time. One goal of Project Prometheus, formerly called the Nuclear Systems Initiative, is to launch a spacecraft towards Jupiter by 2011. The flight would take at least eight years.

The key elements of the HiPEP engine are a high exhaust velocity, a microwave-based method for producing ions that performs for longer than existing technologies and a rectangular design that can more easily be scaled up than circular ones.

Spacecraft are increasingly being built with ion engines rather than engines that burn rocket fuel. This is because ion engines produce more power for a given amount of propellant, and provide a smooth output rather than intermittent spurts.

"Jupiter is such a far away target. Using a chemical system, you just couldn't do it," says John Foster, one of the principal creators of the engine at NASA's Glenn Research Center in Cleveland, Ohio.

The HiPEP engine differs from earlier ion engines, such as that powering NASA's Deep Space One mission, because the xenon ions are produced using a combination of microwaves and spinning magnets. Previously the electrons required were provided by a cathode. Using microwaves significantly reduces the wear and tear on the engine by avoiding any contact between the speeding ions and the electron source.

Nuclear fission

A Japanese asteroid-chasing spacecraft is already using microwave-based technology to produce ions, but Hayabusa uses a small device that could not produce enough power to fly to Jupiter. The HiPEP engine is currently capable of 12 kilowatts of power but its output will be boosted to at least 50 kW for the Jupiter mission.

The rectangular cross section of the HiPEP engine will make this easier, as it can be expanded along one of its sides. A circular engine would have to be rebuilt, says NASA.

Nonetheless, other researchers at NASA's Jet Propulsion Laboratory in Pasadena, California, are working on a cylindrical high-power ion engine, also for the Prometheus project. But Newhouse notes that building a powerful, long-lasting propulsion system is just "one of the pieces we need to get to Jupiter". The electricity for the ion engine is slated to come from on-board nuclear fission reactor. This part of the Prometheus Project is just beginning, with safety considerations, the miniaturization of the reactor and the identity of the fuel all needing to be decided.

NEW IONIC OR BEAM PULSES ENGINES

By this chapter the author proposes a new pulse engine which works with beam or ionic (ionic beam) pulses.

With a new ionic engine one builds a new aircraft (a new ship). The principal characteristic of this kind of engine is the high power (energy) which accelerates the beam at very high energy, in circular accelerators, in modern linear accelerators (LINAC), or in both.

One can use accelerators similar with the static physics accelerators (synchrotron, synchrocyclotron or isochronous cyclotron).

Ionic engine (ion thruster, which accelerates the positive ions through a potential difference) is about 10 times more effective than classic system based on combustion.

We can still improve the efficiency of 10-50 times if one uses positive ions accelerated in a cyclotron mounted on the ship; the efficiency can easily grow for 1000 times if the positive ions will be accelerated in a high energy synchrotron, synchrocyclotron or isochronous cyclotron (1-100 GeV).

Future (ionic) engine will have mandatory a circular particle accelerator (high or very high energy; see the Figure 3).

Sure that the difficulties will arise from design, but they need to be resolved step by step.

We can thus increase the speed and autonomy of the ship using a less quantity of fuel.

One can use synchrotron radiation (synchrotron light, high intensity beams), like high intensity (X-ray or Gamma ray) radiation, as well. In this case will be a beam engine (not an ionic engine).

A linear particle accelerator (also called a LINAC) is an electrical device for the acceleration of subatomic particles. This sort of particle accelerator has many applications. It used recently as to an injector into a higher energy synchrotron at a dedicated experimental particle physics laboratory. In this, the big classic synchrotron is reduced to a ring surface (magnetic core).

The design of a LINAC depends on the type of particle that is being accelerated: electron, proton or ion.

It proposes using a powerful LINAC at the exit of synchrotron (especially when one accelerates electrons) to not lose energy by photons premature emission (figure 3).

One can use a LINAC in the entry in the Synchrotron and one at out (Figure 2). To use a small entrance LINAC, between him and synchrotron, one put an additional speed circuit in a stadium form (Fig. 2).

The end LINAC can be reduced if one put more end LINACs. See diagram below (Fig. 2.).

Fig. 2: *A high energy synchrotron schema*

This ship has two circular particle accelerators (two synchrotrons)

This ship has first a circular particle accelerator (a synchrotron), and at the end two big linear particle accelerators (two big LINAC)

Fig. 3: *Some flying synchrotron prototypes*

CONCLUSIONS

Speaking about a new ionic engine means to speak about a new aircraft.

The chapter presents shortly the actual ionic engines (called ion thrusters) and the new ionic (pulse) engines proposed by the author. Ionic engine (ion thruster, which accelerates the positive ions through a potential difference) is about 10 times more effective than classic system based on combustion.

We can still improve the efficiency of 10-50 times if one uses pulses of positive ions accelerated in a cyclotron mounted on the ship; the efficiency can easily grow for 1000 times if the positive ions will be accelerated in a high energy synchrotron, synchrocyclotron or isochronous cyclotron (1-100 GeV). Future (ionic) engine will have mandatory a circular particle accelerator (high or very high energy). We can thus increase the speed and autonomy of the ship using a less quantity of fuel and power. One can use synchrotron radiation (synchrotron light, high intensity beams), like high intensity (X-ray or Gamma ray) radiation, as well. In this case will be a beam engine (not an ionic engine), it'll use only the power (energy, which can be solar energy, nuclear energy, or both) and so we will remove the fuel.

A linear particle accelerator (also called a LINAC) is an electrical device for the acceleration of subatomic particles. This sort of particle accelerator has many applications. It used recently as to an injector into a higher energy synchrotron at a dedicated experimental particle physics laboratory. In this, the big classic synchrotron is reduced to a ring surface (magnetic core).

The design of a LINAC depends on the type of particle that is being accelerated: electron, proton or ion.

It proposes using a powerful LINAC at the exit of synchrotron (especially when one accelerates electrons) to not lose energy by photons premature emission (figure 3).

One can use a LINAC in the entry in the Synchrotron and one at out (figure 2). To use a small entrance LINAC, between him and synchrotron, one put an additional speed circuit in a stadium form (fig. 2). With a new ionic engine one builds a new aircraft, which can travel through water and. This new aircraft will can accelerate directly, without an additional combustion engine and without gravity assists from other planets.

Ionic engine (ion thruster) has 2 major advantages (a) and 2 disadvantages (b) compared with chemical propulsion; (a) the impulse and energy per unit of fuel used are much higher; 1-the increased impulse generates a higher speed (velocity; so we can walk longer distances in a short time), 2-the high energy decreases fuel consumption and increase the autonomy of the ship; (b) generate force and acceleration are very small; we can't defeat any forces of resistance to lodging by atmosphere and we have no chance to exceed gravitational forces - ship will not leave a planet (or fall on it) using the ion thruster (It required an additional motor). Vacuum ship acceleration is possible but only with very small acceleration. Increasing more the energy (and also the impulse) can reach the necessary forces and acceleration (Growth will need to be very high, 100 PeV-1000 PeV). Particles energy increased can be made with accelerators circular and or modern linear. Particles energy increased will be huge and in addition will need to grow and the flow of accelerated particles (and the tor diameter; if one increases enough the flow, the necessary energy will be 10 GeV-10 TeV).

Immediate consequence of increasing particle energy will be the increasing of speeds and autonomy of the ship. Now we can achieve huge speeds in a very short time. The ship will pass through any atmosphere (including water) with great ease. The ship can take off or land directly.

Initially one can use to ship the old forms (the old design) which adapts and the accelerator(s).

REFERENCES

[1] Wikipedia, *the free encyclopedia*, net,

[2] Dan Tanna, *Technology today*, edit on 10-6-2008, a net Link.

Calculation of the momentum of particle jets

$$m = m_0 \cdot \frac{1}{\sqrt{1 - \dfrac{v^2}{c^2}}} = \frac{m_0 \cdot c}{\sqrt{c^2 - v^2}} \quad Lorentz$$

$$\frac{dm}{dv} = \frac{m \cdot v}{c^2 - v^2}$$

$$E_C = \frac{1}{2} \cdot m \cdot v^2$$

$$p = \frac{dE_c}{dv} = \frac{dm}{dv} \cdot \frac{v^2}{2} + \frac{m}{2} \cdot \frac{dv^2}{dv} = \frac{m \cdot v \cdot (2 \cdot c^2 - v^2)}{(2 \cdot c^2 - 2 \cdot v^2)} \Rightarrow$$

$$\Rightarrow p = \frac{m_0 \cdot c \cdot v \cdot (2 \cdot c^2 - v^2)}{2 \cdot \sqrt{c^2 - v^2} \cdot (c^2 - v^2)} = \frac{m_0 \cdot c \cdot v \cdot (2 \cdot c^2 - v^2)}{2 \cdot (c^2 - v^2)^{3/2}}$$

$$\Rightarrow$$

$$\Rightarrow \begin{cases} p = \dfrac{m_0 \cdot c \cdot v \cdot (2 \cdot c^2 - v^2)}{2 \cdot (c^2 - v^2)^{3/2}} \quad when \quad v \neq c \\[4mm] p = \dfrac{h}{\lambda} \quad when \quad v \equiv c \end{cases}$$

$\Rightarrow k = M \cdot n \cdot N \cdot p$ Momentum   of particle   jets

n = Number of pulses per second

N = average number of particles per pulse

M = number of ship engines

$$V_s \cdot M_s = k \Rightarrow V_s = \frac{k}{M_s} \begin{cases} V_s = \text{the speed of the ship} \\ M_s = \text{the mass of the ship} \end{cases}$$

Calculation of the acceleration of the ship

$$
\begin{cases}
\dfrac{dp}{dt} = \dfrac{3\,p\cdot(c^2 - v^2)^{1/2}\cdot v + m_0\cdot c\cdot\left(c^2 - \dfrac{3}{2}v^2\right)}{(c^2 - v^2)^{3/2}}\cdot\dfrac{dv}{dt} \quad \text{when} \quad v \neq c \\[4mm]
\dfrac{dp}{dt} = \dfrac{h}{c}\cdot\dfrac{dv}{dt} \quad \text{when} \quad v \equiv c
\end{cases}
$$

$$
\begin{cases}
F = \dfrac{dk}{dt} = M\cdot n\cdot N\cdot\dfrac{dp}{dt} \;\Rightarrow\; a_s = \dfrac{M\cdot n\cdot N}{M_s}\cdot\dfrac{dp}{dt} \\[4mm]
F = M_s\cdot a_s
\end{cases}
$$

Condition of the ship output from Earth to space

$$
a_s \geq g \;\Rightarrow\; \dfrac{M\cdot n\cdot N}{M_s}\cdot\dfrac{dp}{dt} \geq g
$$

BIBLIOGRAPHY

[1] "Japan Airlines announces delivery of 787". Japan Airlines. March 21, 2012. Retrieved June 12, 2012.

[2] Gates, Dominic (September 24, 2011). "Boeing celebrates 787 delivery as program's costs top $32 billion". The Seattle Times. Retrieved September 26, 2011.

[3] "Commercial Airplanes – 787 Dreamliner – Background" (Press release). Boeing. Retrieved December 14, 2010.

[4] "Flight Standardization Board Report". Federal Aviation Administration. August 25, 2011. Retrieved November 8, 2011.

[5] "Boeing Gives the 7E7 Dreamliner a Model Designation" (Press release). Boeing. January 28, 2005. Retrieved June 14, 2011.

[6] "Boeing Celebrates the Premiere of the 787 Dreamliner" (Press release). Boeing. July 8, 2007. Retrieved January 21, 2011.

[7] Gunter, Lori (July 2002). "The Need for Speed, Boeing's Sonic Cruiser team focuses on the future". Boeing Frontier magazine. Retrieved January 21, 2011.

[8] Banks, Howard (May 28, 2001). "Paper plane: That Mach 0.95 Sonic Cruiser from Boeing will never fly. Here's why.". Forbes. Retrieved June 7, 2007.

[9] Associated Press. "History of the Boeing 787".[dead link] ABC News, June 23, 2009. Retrieved June 23, 2009.

[10] Cannegieter, Roger. "Long Range vs. Ultra High Capacity". Aerlines.nl. Retrieved April 8, 2010.

[11] Babej, Marc E.; Pollak, Tim (May 24, 2006). "Boeing Versus Airbus". Forbes. Retrieved April 8, 2010.

[12] "Maximizing the Middle, Finding the sweet spot in the market" (Press release). Boeing Frontier magazine. March 2003.

[13] "Boeing Achieves 787 Power On" (Press release). Boeing. June 20, 200.

[14] "'Name Your Plane' sweepstakes". Boeing Frontiers Online. July 2003. Retrieved September 28, 2007.

[15] "Boeing Launches 7E7 Dreamliner" (Press release). Boeing. April 26, 2004. Retrieved June 14, 2011.

[16] Ogando, Joseph (June 7, 2007). "Design News – Features – Boeing's 'More Electric' 787 Dreamliner Spurs Engine Evolution". designnews.com. Retrieved September 7, 2011.

[17] Pandey, Mohan (2010). How Boeing Defied the Airbus Challenge. USA: Createspace. ISBN 978-1-4505-0113-2.

[18] "Boeing Unveils 787 Final Assembly Factory Flow." Boeing, December 6, 2006. Retrieved September 2011 3, 2011.

[19] "Boeing's Big Dream", Fortune Magazine, May 5, 2008, p. 182. (online version).

[20] "Boeing's Big Dream", Fortune, May 5, 2008, p. 184.

[21] Seo, Sookyung (September 29, 2010). "Boeing 787 Supplier Korea Aerospace Hires Share-Sale Arrangers". Bloomberg. Retrieved September 2, 2011.

[22] "Boeing Still Working On 787 Weight Issue, Carson Says". Associated Press. December 7, 2006.

[23] "Boeing to deliver test 787s to its customers". Financial Times. July 6, 2007.

[24] Sanders, Peter (July 8, 2009). "Boeing Sets Deal to Buy a Dreamliner Plant". Wall Street Journal.

[25] "Boeing Confirms Success on 787 Wing, Fuselage Ultimate Load Test" (Press release). Boeing. April 7, 2010.

[26] "Boeing Testing Sample Sonic Cruiser Fuselage". Boeing. July 24, 2002.

[27] "Development Work on Boeing 787 Noses Ahead". Boeing. July 13, 2005. Retrieved June 14, 2011.

[28] "Orders and Deliveries search page". Boeing. July 21, 2010.

[29] Congressional Research Service (1992). Airbus Industrie: An Economic and Trade Perspective. U.S. Library of Congress.

[30] Heppenheimer, T.A. (1995). Turbulent Skies: The History of Commercial Aviation. John Wiley. ISBN 0-471-19694-0.

[31] Lynn, Matthew (1997). Birds of Prey: Boeing vs. Airbus, a Battle for the Skies. Four Walls Eight Windows. ISBN 1-56858-107-6.

[32] McGuire, Steven (1997). Airbus Industrie: Conflict and Cooperation in U.S.E.C. Trade Relations. St. Martin's Press.

[33] McIntyre, Ian (1982). Dogfight: The Transatlantic Battle Over Airbus. Praeger Publishers. ISBN 0-275-94278-3.

[34] Thornton, David Weldon (1995). Airbus Industrie: The Politics of an International Industrial Collaboration. St. Martin's Press. ISBN 0-312-12441-4.

[35] Coox, Alvin D. Nomonhan: Japan Against Russia, 1939. Two volumes; 1985, Stanford University Press. ISBN 0-8047-1160-7.

[36] Eden, Paul (General Editor). The Encyclopedia of Aircraft of WWII. London: Amber Books, Ltd, 2004. ISBN 1-904687-07-5.

[37] Ireland, Bernard and Eric Grove. Jane's War at Sea. UK: Harper Collins Publishers, 1997. ISBN 0-00-472065-2.

[38] Mark (1995-07). Aerial Interdiction: Air Power and the Land Battle in Three American Wars. pp. 9–10. ISBN 978-0-7881-1966-8.

[39] Brown, Michael E. Flying Blind: The Politics of the U.S. Strategic Bomber Program. Ithaca, NY: Cornell University Press, 1992.

[40] Cross, Robin. The Bombers: The Illustrated Story of Offensive Strategy and Tactics in the Twentieth Century. New York: Macmillan, 1987.

[41] Green, William. Famous Bombers of the Second World War. New York: Doubleday, 1959, 1960 (two vols).

[42] Green, William. Warplanes of the Third Reich. New York: Doubleday, 1970.

[43] Haddow, G. W., and Peter M. Grosz The German Giants: The German R-Planes 1914-1918. London: Putnam, 1969 (2nd ed.).

[44] Hastings, Max. Bomber Command. New York: Dial Press, 1979.

[45] Jones, Lloyd S. U.S. Bombers 1926 to 1980s. Fallbrook, CA: Aero Publishers, 1980 (3rd ed.).

[46] Neillands, Robin. The Bomber War: The Allied Offensive Against Nazi Germany. Woodstock, NY: Overlook, 2001.

[47] Robinson, Douglas H. The Zeppelin in Combat: A History of the German Naval Airship Division, 1912-1918. Atglen, PA: Schiffer, 1994.

[48] United States Strategic Bombing Survey. Over-all Report (European War). Washington: Government Printing Office, September 30, 1945.

[49] Lee, Arthur Gould. No Parachute. London: Jarrolds, 1968. ISBN 0-09-086590-1.

[50] Eric Lawson, Jane Lawson (2002). "The First Air Campaign: August 1914–November 1918". Da Capo Press. p.123. ISBN 0-306-81213-4.

[51] Harry Furniss (2000). "Memoirs one: the flying game". Trafford Publishing. ISBN 1-55212-513-0.

[52] John Buckley (1998). "Air power in the age of total war". Taylor & Francis. p.43. ISBN 1-85728-589-1.

[53] Munson, Kenneth. Fighters and Bombers of World War II. New York City: Peerage Books, 1983, p. 159. ISBN 0-907408-37-0.

[54] Office Of Air Force History Washingtondc, Daniel R Mortensen. A Pattern for Joint Operations: World War II Close Air Support, North Africa. pp. 24–25. ISBN 978-1-4289-1564-0.

[55] E. R. Johnson (2008-08). American Attack Aircraft Since 1926. p. 413. ISBN 978-0-7864-3464-0. Retrieved 21 January 2011.

[56] Merriman, Ray (2000). U.S. warplanes of World War II, volume 1. Bennington, VT: Merriam Press. p. 3. ISBN 978-1-57638-167-0. Retrieved 21 January 2011. ""A: Light Bombing [...] B: Medium and Heavy Bombing"".

[57] Corum, James S; Wray R Johnson (2003). Airpower in Small Wars — Fighting Insurgents and Terrorists. Lawrence, Kansas: University Press of Kansas. pp. 23–40. ISBN 0-7006-1240-8.

[58] Hallion, Richard (2010-10-28). Strike from the Sky: The History of Battlefield Air Attack, 1910-1945. pp. 3–6. ISBN 978-0-8173-5657-6.

[59] Franklin Cooling, B; Office Of Air Force History, United States. Air Force (1990). Case studies in the development of close air support. pp. 101, 123. ISBN 978-0-912799-64-3.

[60] Ian Gooderson (1998). Air power at the battlefront: allied close air support in Europe, 1943-45. Routledge. pp. 121. ISBN 978-0-7146-4680-0.

[61] Shores, Christopher and Thomas, Chris. Second Tactical Air Force Volume Two. Breakout to Bodenplatte July 1944 to January 1945. Hersham, Surrey, UK: Ian Allan Publishing Ltd, 2005. ISBN 1-903223-41-5, pages 245-250.

[62] From Hot Air to Hellfire, James W. Bradin, ISBN 0-89141-511-4.

[63] "Inside story of the troubled F/A-18." Popular Science, Volume 223, Issue 4, October 1983. ISSN 0161-73702. Retrieved: 23 December 2011.

[64] "The FY 1981 military programs." Bulletin of the Atomic Scientists, Volume 36, Issue 6, June 1980, p. 38. ISSN 0096-3402. Retrieved: 23 December 2011.